Foundations of Plasma Physics

Foundations of Plasma Physics

Eric Buchanan

Ⓜ MURPHY & MOORE
www.murphy-moorepublishing.com

Published by Murphy & Moore Publishing
1 Rockefeller Plaza,
New York City, NY 10020, USA
www.murphy-moorepublishing.com

Foundations of Plasma Physics
Eric Buchanan

International Standard Book Number: 978-1-63987-237-4 (Hardback)

Cataloging-in-Publication Data

Foundations of plasma physics / Eric Buchanan.
p. cm.
Includes bibliographical references and index.
ISBN 978-1-63987-237-4
1. Plasma (Ionized gases). 2. Ionized gases. 3. Physics. I. Buchanan, Eric.
QC718 .P53 2022
530.44--dc23

TABLE OF CONTENTS

The purpose of this book is to help students understand the fundamental concepts of this discipline. It is designed to motivate students to learn and prosper. I am grateful for the support of my colleagues. I would also like to acknowledge the encouragement of my family.

In physics, plasma is the fourth state of matter, the others being solid, gas and liquid. It is an electrically neutral medium of untied positive and negative particles. Plasma consists of gas of ions and some of their orbital electrons are removed. Plasma physics is a sub-discipline of physics which is focused on the study of plasma. Bulk interactions, plasma approximation, and plasma frequency are the key factors that define plasma. They are great electrical conductors. Plasma modeling is the solving of equations related to motion which describe the state of plasma. Artificial plasma can be generated by applying electric and magnetic fields through a gas. Plasma stability is an important concept in the field of plasma physics. It determines whether the perturbation will grow further, oscillate or damped out. This book attempts to understand the multiple branches that fall under the discipline of plasma physics and how such concepts have practical applications. Such selected concepts that redefine this field have been presented in it. This textbook aims to serve as a resource guide for students and experts alike and contribute to the growth of the discipline.

A foreword for all the chapters is provided below:

Chapter – Introduction

A plasma is a hot ionized gas consisting of approximately equal numbers of positively charged ions and negatively charged electrons. It is one of the four fundamental states of matter. This is an introductory chapter which will briefly introduce all the significant aspects of plasma physics.

Chapter – Significant Aspects of Plasma Physics

Some of the fundamental concepts that are studied under plasma physics are surface-wave-sustained discharge, plasma stability, Thomson scattering, plasma parameters, corona discharge, coronal seismology, diffusion damping, double layer, etc. This chapter closely examines these fundamental concepts of plasma physics to provide an extensive understanding of the subject.

Chapter – Waves in Plasmas

Waves in plasmas refer to an interconnected set of particles and fields that propagate periodically. Electromagnetic electron wave, ion acoustic wave, and Alfvén wave are some of the examples of these waves. Plasma oscillation, upper and lower hybrid oscillation are also studied under it. The topics elaborated in this chapter will help in gaining a better perspective about these waves in plasma.

Chapter – Plasma Diagnostics

Plasma diagnostics refers to the methods, techniques and instruments that are used for the measurement of varied properties of plasma. Some of the most common invasive probe methods used are Langmuir probe, ball-pen probe and Faraday cup. The chapter closely examines these key concepts of plasma diagnostics to provide an extensive understanding of the subject.

Chapter – Plasma Processing

The plasma-based material processing technology which modifies the chemical and physical properties of a surface is referred to as plasma processing. A few of the techniques include plasma surface activation, plasma ashing, plasma cleaning, corona treatment, plasma etching, plasma functionalization, plasma polymerization, etc. All these diverse techniques of plasma processing have been carefully analyzed in this chapter.

Eric Buchanan

Introduction

A plasma is a hot ionized gas consisting of approximately equal numbers of positively charged ions and negatively charged electrons. It is one of the four fundamental states of matter. This is an introductory chapter which will briefly introduce all the significant aspects of plasma physics.

PLASMA

Plasma, in physics is an electrically conducting medium in which there are roughly equal numbers of positively and negatively charged particles, produced when the atoms in a gas become ionized. It is sometimes referred to as the fourth state of matter, distinct from the solid, liquid, and gaseous states.

The negative charge is usually carried by electrons, each of which has one unit of negative charge. The positive charge is typically carried by atoms or molecules that are missing those same electrons. In some rare but interesting cases, electrons missing from one type of atom or molecule become attached to another component, resulting in a plasma containing both positive and negative ions. The most extreme case of this type occurs when small but macroscopic dust particles become charged in a state referred to as a dusty plasma. The uniqueness of the plasma state is due to the importance of electric and magnetic forces that act on a plasma in addition to such forces as gravity that affect all forms of matter. Since these electromagnetic forces can act at large distances, a plasma will act collectively much like a fluid even when the particles seldom collide with one another.

Nearly all the visible matter in the universe exists in the plasma state, occurring predominantly in this form in the Sun and stars and in interplanetary and interstellar space. Auroras, lightning, and welding arcs are also plasmas; plasmas exist in neon and fluorescent tubes, in the crystal structure of metallic solids, and in many other phenomena and objects. The Earth itself is immersed in a tenuous plasma called the solar wind and is surrounded by a dense plasma called the ionosphere.

A plasma may be produced in the laboratory by heating a gas to an extremely high temperature, which causes such vigorous collisions between its atoms and molecules that electrons are ripped free, yielding the requisite electrons and ions. A similar process occurs inside stars. In space the dominant plasma formation process is photoionization, wherein photons from sunlight or starlight are absorbed by an existing gas, causing electrons to be emitted. Since the Sun and stars shine continuously, virtually all the matter becomes ionized in such cases, and the plasma is said to be fully ionized. This need not be the case, however, for a plasma may be only partially ionized. A completely ionized hydrogen plasma, consisting solely of electrons and protons (hydrogen nuclei), is the most elementary plasma.

Plasma Physics

The modern concept of the plasma state is of recent origin, dating back only to the early 1950s. Its history is interwoven with many disciplines. Three basic fields of study made unique early contributions to the development of plasma physics as a discipline: electric discharges, magnetohydrodynamics (in which a conducting fluid such as mercury is studied), and kinetic theory.

Interest in electric-discharge phenomena may be traced back to the beginning of the 18th century, with three English physicists—Michael Faraday in the 1830s and Joseph John Thomson and John Sealy Edward Townsend at the turn of the 19th century—laying the foundations of the present understanding of the phenomena. Irving Langmuir introduced the term plasma in 1923 while investigating electric discharges. In 1929 he and Lewi Tonks, another physicist working in the United States, used the term to designate those regions of a discharge in which certain periodic variations of the negatively charged electrons could occur. They called these oscillations plasma oscillations, their behaviour suggesting that of a jellylike substance. Not until 1952, however, when two other American physicists, David Bohm and David Pines, first considered the collective behaviour of electrons in metals as distinct from that in ionized gases, was the general applicability of the concept of a plasma fully appreciated.

The collective behaviour of charged particles in magnetic fields and the concept of a conducting fluid are implicit in magneto hydrodynamic studies, the foundations of which were laid in the early and middle 1800s by Faraday and André-Marie Ampère of France. Not until the 1930s, however, when new solar and geophysical phenomena were being discovered, were many of the basic problems of the mutual interaction between ionized gases and magnetic fields considered. In 1942 Hannes Alfvén, a Swedish physicist, introduced the concept of magneto hydrodynamic waves. This contribution, along with his further studies of space plasmas, led to Alfvén's receipt of the Nobel Prize for Physics in 1970.

These two separate approaches—the study of electric discharges and the study of the behaviour of conducting fluids in magnetic fields—were unified by the introduction of the kinetic theory of the plasma state. This theory states that plasma, like gas, consists of particles in random motion, whose interactions can be through long-range electromagnetic forces as well as via collisions. In 1905 the Dutch physicist Hendrik Antoon

Lorentz applied the kinetic equation for atoms (the formulation by the Austrian physicist Ludwig Eduard Boltzmann) to the behaviour of electrons in metals. Various physicists and mathematicians in the 1930s and '40s further developed the plasma kinetic theory to a high degree of sophistication. Since the early 1950s interest has increasingly focused on the plasma state itself. Space exploration, the development of electronic devices, a growing awareness of the importance of magnetic fields in astrophysical phenomena, and the quest for controlled thermonuclear (nuclear fusion) power reactors all have stimulated such interest. Many problems remain unsolved in space plasma physics research, owing to the complexity of the phenomena. For example, descriptions of the solar wind must include not only equations dealing with the effects of gravity, temperature, and pressure as needed in atmospheric science but also the equations of the Scottish physicist James Clerk Maxwell, which are needed to describe the electromagnetic field.

Plasma Formation

Apart from solid-state plasmas, such as those in metallic crystals, plasmas do not usually occur naturally at the surface of the Earth. For laboratory experiments and technological applications, plasmas therefore must be produced artificially. Because the atoms of such alkalies as potassium, sodium, and cesium possess low ionization energies, plasmas may be produced from these by the direct application of heat at temperatures of about 3,000 K. In most gases, however, before any significant degree of ionization is achieved, temperatures in the neighbourhood of 10,000 K are required. A convenient unit for measuring temperature in the study of plasmas is the electron volt (eV), which is the energy gained by an electron in vacuum when it is accelerated across one volt of electric potential. The temperature, W, measured in electron volts is given by $W = T/12,000$ when T is expressed in kelvins. The temperatures required for self-ionization thus range from 2.5 to 8 electron volts, since such values are typical of the energy needed to remove one electron from an atom or molecule.

Because all substances melt at temperatures far below that level, no container yet built can withstand an external application of the heat necessary to form a plasma; therefore, any heating must be supplied internally. One technique is to apply an electric field to the gas to accelerate and scatter any free electrons, thereby heating the plasma. This type of ohmic heating is similar to the method in which free electrons in the heating element of an electric oven heat the coil. Because of their small energy loss in elastic collisions, electrons can be raised to much higher temperatures than other particles. For plasma formation a sufficiently high electric field must be applied, its exact value depending on geometry and the gas pressure. The electric field may be set up via electrodes or by transformer action, in which the electric field is induced by a changing magnetic field. Laboratory temperatures of about 10,000,000 K, or 8 kiloelectron volts (keV), with electron densities of about 10^{19} per cubic metre have been achieved by the transformer method. The temperature is eventually limited by energy losses to the outside environment. Extremely high temperatures, but relatively low-density plasmas, have been produced by the separate injection of ions and electrons into a mirror system (a plasma device using a particular arrangement of magnetic fields for containment). Other methods have used the high temperatures that develop behind a wave that is moving much faster than sound to produce what is called a shock front; lasers have also been employed.

Natural plasma heating and ionization occur in analogous ways. In a lightning-induced plasma, the electric current carried by the stroke heats the atmosphere in the same manner as in the ohmic heating technique described above. In solar and stellar plasmas the heating is internal and caused by nuclear fusion reactions. In the solar corona, the heating occurs because of waves that propagate from the surface into the Sun's atmosphere, heating the plasma much like shock-wave heating in laboratory plasmas. In the ionosphere, ionization is accomplished not through heating of the plasma but rather by the flux of energetic photons from the Sun. Far-ultraviolet rays and X rays from the Sun have enough energy to ionize atoms in the Earth's atmosphere. Some of the energy also goes into heating the gas, with the result that the upper atmosphere, called the thermosphere, is quite hot. These processes protect the Earth from energetic photons much as the ozone layer protects terrestrial life-forms from lower-energy ultraviolet light. The typical temperature 300 kilometres above the Earth's surface is 1,200 K, or about 0.1 eV. Although it is quite warm compared with the surface of the Earth, this temperature is too low to create self-ionization. When the Sun sets with respect to the ionosphere, the source of ionization ceases, and the lower portion of the ionosphere reverts to its nonplasma state. Some ions, in particular singly charged oxygen (O^+), live long enough that some plasma remains until the next sunrise. In the case of an aurora, a plasma is created in the nighttime or daytime atmosphere when beams of electrons are accelerated to hundreds or thousands of electron volts and smash into the atmosphere.

Methods of describing Plasma Phenomena

The behaviour of a plasma may be described at different levels. If collisions are relatively infrequent, it is useful to consider the motions of individual particles. In most plasmas of interest, a magnetic field exerts a force on a charged particle only if the particle is moving, the force being at right angles to both the direction of the field and the direction of particle motion. In a uniform magnetic field (B), a charged particle gyrates about a line of force. The centre of the orbit is called the guiding centre. The particle may also have a component of velocity parallel to the magnetic field and so traces out a helix in a uniform magnetic field. If a uniform electric field (E) is applied at right angles to the direction of the magnetic field, the guiding centre drifts with a uniform velocity of magnitude equal to the ratio of the electric to the magnetic field (E/B), at right angles to both the electric and magnetic fields. A particle starting from rest in such fields follows the same cycloidal path a dot on the rim of a rolling wheel follows. Although the "wheel" radius and its sense of rotation vary for different particles, the guiding centre moves at the same E/B velocity, independent of the particle's charge and mass. Should the electric field change with time, the problem would become even more complex. If, however, such an alternating electric field varies at the same frequency as the cyclotron frequency (i.e., the rate of gyration), the guiding centre will remain stationary, and the particle will be forced to travel in an ever-expanding orbit. This phenomenon is called cyclotron resonance and is the basis of the cyclotron particle accelerator.

The motion of a particle about its guiding centre constitutes a circular current. As such, the motion produces a dipole magnetic field not unlike that produced by a simple bar magnet. Thus, a moving charge not only interacts with magnetic fields but also produces them. The direction of the magnetic field produced by a moving particle, however, depends both on whether the particle is positively or negatively charged and on the direction of its motion. If the motion of the charged particles is completely random, the net associated magnetic field is zero. On the other hand, if charges of different sign have an average relative velocity (i.e., if an electric current flows), then a net magnetic field over and above any externally applied field exists. The magnetic interaction between charged particles is therefore of a collective, rather than of an individual, particle nature.

At a higher level of description than that of the single particle, kinetic equations of the Boltzmann type are used. Such equations essentially describe the behaviour of those particles about a point in a small-volume element, the particle velocities lying within a small range about a given value. The interactions with all other velocity groups, volume elements, and any externally applied electric and magnetic fields are taken into account. In many cases, equations of a fluid type may be derived from the kinetic equations; they express the conservation of mass, momentum, and energy per unit volume, with one such set of equations for each particle type.

Determination of Plasma Variables

The basic variables useful in the study of plasma are number densities, temperatures, electric and magnetic field strengths, and particle velocities. In the laboratory and in space, both electrostatic (charged) and magnetic types of sensory devices called probes help determine the magnitudes of such variables. With the electrostatic probe, ion densities, electron and ion temperatures, and electrostatic potential differences can be determined. Small search coils and other types of magnetic probes yield values for the magnetic field; and from Maxwell's electromagnetic equations the current and charge densities and the induced component of the electric field may be found. Interplanetary spacecraft have carried such probes to nearly every planet in the solar system, revealing to scientists such plasma phenomena as lightning on Jupiter and the sounds of Saturn's rings and radiation belts. In the early 1990s, signals were being relayed to the Earth from several spacecraft approaching the edge of the plasma boundary to the solar system, the heliopause.

In the laboratory the absorption, scattering, and excitation of neutral and high-energy ion beams are helpful in determining electron temperatures and densities; in general, the refraction, reflection, absorption, scattering, and interference of electromagnetic waves also provide ways to determine these same variables. This technique has also been employed to remotely measure the properties of the plasmas in the near-space regions of the Earth using the incoherent scatter radar method. The largest single antenna is at the National Astronomy and Ionosphere Center at Arecibo in Puerto Rico. It has a circumference of 305 metres and was completed in 1963. It is still used to probe space plasmas to distances of 3,000 kilometres. The method works by bouncing radio waves from small irregularities in the electron gas that occur owing to random thermal motions of the particles. The returning signal is shifted slightly from the transmitted one—because of the Doppler-shift effect—and the velocity of the plasma can be determined in a manner similar to the way in which the police detect a speeding car. Using this method, the wind speed in space can be found, along with the temperature, density, electric field, and even the types of ions present. In geospace the appropriate radar frequencies are in the range of 50 to 1,000 megahertz (MHz), while in the laboratory, where the plasma densities and plasma frequencies are higher, microwaves and lasers must be used.

Aside from the above methods, much can be learned from the radiation generated and emitted by the plasma itself; in fact, this is the only means of studying cosmic plasma beyond the solar system. The various spectroscopic techniques covering the entire continuous radiation spectrum determine temperatures and identify such nonthermal sources as those pulses producing synchrotron radiations.

Low-frequency Waves

At the lowest frequency are Alfvén waves, which require the presence of a magnetic field to exist. In fact, except for ion acoustic waves, the existence of a background magnetic field is required for any wave with a frequency less than the plasma frequency to occur in a plasma. Most natural plasmas are threaded by a magnetic field, and laboratory plasmas often use a magnetic field for confinement, so this requirement is usually met, and all types of waves can occur.

Alfvén waves are analogous to the waves that occur on the stretched string of a guitar. In this case, the string represents a magnetic field line. When a small magnetic field disturbance takes place, the field is bent slightly, and the disturbance propagates in the direction of the magnetic field. Since any changing magnetic field creates an electric field, an electromagnetic wave results. Such waves are the slowest and have the lowest frequencies of any known electromagnetic waves. For example, the solar wind streams out from the Sun with a speed greater than either electromagnetic (Alfvén) or sound waves. This means that, when the solar wind hits the Earth's outermost magnetic field lines, a shock wave results to "inform" the incoming plasma that an obstacle exists, much like the shock wave associated with a supersonic airplane. The shock wave travels toward the Sun at the same speed but in the opposite direction as the solar wind, so it appears to stand still with respect to the Earth. Because there are almost no particle-particle collisions, this type of collisionless shock wave is of great interest to space plasma physicists who postulate that similar shocks occur around supernovas and in other astrophysical plasmas. On the Earth's side of the shock wave, the heated and slowed solar wind interacts with the Earth's atmosphere via Alfvén waves propagating along the magnetic field lines.

The turbulent surface of the Sun radiates large-amplitude Alfvén waves, which are thought to be responsible for heating the corona to 1,000,000 K. Such waves can also produce fluctuations in the solar wind, and, as they propagate through it to the Earth, they seem to control the occurrence of magnetic storms and auroras that are capable of disrupting communication systems and power grids on the planet.

Two fundamental types of wave motion can occur: longitudinal, like a sound or ion acoustic wave, in which particle oscillation is in a direction parallel to the direction of wave propagation; and transverse, like a surface water wave, in which particle oscillation is in a plane perpendicular to the direction of wave propagation. In all cases, a wave may be characterized by a speed of propagation (u), a wavelength (λ), and a frequency (v) related by an expression in which the velocity is equal to the product of the wavelength and frequency, namely, $u = \lambda v$. The Alfvén wave is a transverse wave and propagates with a velocity that depends on the particle density and the magnetic field strength. The velocity is equal to the magnetic flux density (B) divided by the square root of the mass density (ρ) times the permeability of free space (μ_o)—that is to say, $B/$ Square root of$\sqrt{\mu_o \rho}$. The ion acoustic wave is a longitudinal wave and also propagates parallel to the magnetic field at a speed roughly equal to the average thermal velocity of the ions. Perpendicular to the magnetic field a different type of longitudinal wave called a magnetosonic wave can occur.

Higher Frequency Waves

In these waves the plasma behaves as a whole, and the velocity is independent of wave frequency. At higher frequencies, however, the separate behaviour of ions and electrons causes the wave velocities to vary with direction and frequency. The Alfvén wave splits into two components, referred to as the fast and slow Alfvén waves, which propagate at different frequency-dependent speeds. At still higher frequencies these two waves (called the electron cyclotron and ion cyclotron waves, respectively) cause electron and cyclotron resonances(synchronization) at the appropriate resonance frequencies. Beyond these resonances, transverse wave propagation does not occur at all until frequencies comparable to and above the plasma frequency are reached.

At frequencies between the ion and electron gyro frequencies lies a wave mode called a whistler. This name comes from the study of plasma waves generated by lightning. When early researchers listened to natural radio waves by attaching an antenna to an audio amplifier, they heard a strange whistling sound. The whistle occurs when the electrical signal from lightning in one hemisphere travels along the Earth's magnetic field lines to the other hemisphere. The trip is so long that some waves (those at higher frequencies) arrive first, resulting in the generation of a whistle like sound. These natural waves were used to probe the region of space around the Earth before spacecraft became available. Such a frequency-dependent wave velocity is called wave dispersion because the various frequencies disperse with distance.

The speed of an ion acoustic wave also becomes dispersive at high frequencies, and a resonance similar to electron plasma oscillations occurs at a frequency determined by electrostatic oscillations of the ions. Beyond this frequency no sonic wave propagates parallel to a magnetic field until the frequency reaches the plasma frequency, above which electroacoustic waves occur. The wavelength of these waves at the critical frequency (ωp) is infinite, the electron behaviour at this frequency taking the form of the plasma oscillations of Langmuir and Tonks. Even without particle collisions, waves shorter than the Debye length are heavily damped—i.e., their amplitude decreases rapidly with time. This phenomenon, called Landau damping, arises because some electrons have the same velocity as the wave. As they move with the wave, they are accelerated much like a surfer on a water wave and thus extract energy from the wave, damping it in the process.

Containment

Magnetic fields are used to contain high-density, high-temperature plasmas because such fields exert pressures and tensile forces on the plasma. An equilibrium configuration is reached only when at all points in the plasma these pressures and tensions exactly balance the pressure from the motion of the particles. A well-known example of this is the effect observed in specially designed equipment. If an external electric current is imposed on a cylindrically shaped plasma and flows parallel to the plasma axis, the magnetic forces act inward and cause the plasma to constrict, or pinch. An equilibrium condition is reached in which the temperature is proportional to the square of the electric current. This result suggests that any temperature may be achieved by making the electric current sufficiently large, the heating resulting from currents and compression. In practice, however, since no plasma can be infinitely long, serious energy losses occur at the ends of the cylinder; also, major instabilities develop in such a simple configuration. Suppression of such instabilities has been one of the major efforts in laboratory plasma physics and in the quest to control the fusion reaction.

A useful way of describing the confinement of a plasma by a magnetic field is by measuring containment time (τc), or the average time for a charged particle to diffuse out of the plasma; this time is different for each type of configuration. Various types of instabilities can occur in plasma. These lead to a loss of plasma and a catastrophic decrease in containment time. The most important of these is called magneto hydrodynamic instability. Although an equilibrium state may exist, it may not correspond to the lowest possible energy. The plasma, therefore, seeks a state of lower potential energy, just as a ball at rest on top of a hill (representing an equilibrium state) rolls down to the bottom if perturbed; the lower energy state of the plasma corresponds to a ball at the bottom of a valley. In seeking the lower energy state, turbulence develops, leading to enhanced diffusion, increased electrical resistivity, and large heat losses. In toroidal geometry, circular plasma currents must be kept below a critical value called the Kruskal-Shafranov limit, otherwise a particularly violent instability consisting of a series of kinks may occur. Although a completely stable system appears to be virtually impossible, considerable progress has been made in devising systems that eliminate the major instabilities. Temperatures on the order of 10,000,000 K at densities of 10^{19} particles per cubic metre and containment times as high as 1/50 of a second have been achieved.

Applications of Plasmas

The most important practical applications of plasmas lie in the future, largely in the field of power production. The major method of generating electric power has been to use heatsources to convert water to steam, which drives turbogenerators. Such heat sources depend on the combustion of fossil fuels, such as coal, oil, and natural gas, and fission processes in nuclear reactors. A potential source of heat might be supplied by a fusion reactor, with a basic element of deuterium-tritium plasma; nuclear fusion collisions between those isotopes of hydrogen would release large amounts of energy to the kinetic energy of the reaction products (the neutrons and the nuclei of hydrogen and helium atoms). By absorbing those products in a surrounding medium, a powerful heat source could be created. To realize a net power output from such a generating station—allowing for plasma radiation and particle losses and for the somewhat inefficient conversion of heat to electricity—plasma temperatures of about 10,000,000 K and a product of particle density times containment time of about 10^{20} seconds per cubic metre are necessary. For example, at a density of 10^{20} particles per metre cubed, the containment time must be one second. Such figures are yet to be reached, although there has been much progress.

In general, there are two basic methods of eliminating or minimizing end losses from an artificially created plasma: the production of toroidal plasmas and the use of magnetic mirrors. A toroidal plasma is essentially one in which a plasma of cylindrical cross section is bent in a circle so as to close on itself. For such plasmas to be in equilibrium and stable, however, special magnetic fields are required, the largest component of which is a circular field parallel to the axis of the plasma. In addition, a number of turbulent plasma processes must be controlled to keep the system stable. In 1991 a machine called the JET (Joint European Torus) was able to generate 1.7 million watts of fusion power for almost 2 seconds after researchers injected titrium into the JET's magnetically confined plasma. It was the first successful controlled production of fusion power in such a confined medium.

Besides generating power, a fusion reactor might desalinate seawater. Approximately two-thirds of the world's land surface is uninhabited, with one-half of this area being arid. The use of both giant fission and fusion reactors in the large-scale evaporation of seawater could make irrigation of such areas economically feasible. Another possibility in power production is the elimination of the heat–steam–mechanical energy chain. One suggestion depends on the dynamo effect. If a plasma moves perpendicular to a magnetic field, an electromotive force, according to Faraday's law, is generated in a direction perpendicular to both the direction of flow of the plasma and the magnetic field. This dynamo effect can drive a current in an external circuit connected to electrodes in the plasma, and thus electric power may be produced without the need for steam-driven rotating machinery. This process is referred to as magneto hydrodynamic (MHD) power generation and has been proposed as a method of extracting power from certain types of fission reactors. Such a generator powers the auroras as the Earth's magnetic field lines tap electrical current from the MHD generator in the solar wind.

The inverse of the dynamo effect, called the motor effect, may be used to accelerate plasma. By pulsing cusp-shaped magnetic fields in a plasma, for example, it is possible to achieve thrusts proportional to the square of the magnetic field. Motors based on such a technique have been proposed for the propulsion of craft in deep space. They have the advantage of being capable of achieving large exhaust velocities, thus minimizing the amount of fuel carried.

A practical application of plasma involves the glow discharge that occurs between two electrodes at pressures of one-thousandth of an atmosphere or thereabouts. Such glow discharges are responsible for the light given off by neon tubes and such other light sources as fluorescent lamps, which operate by virtue of the plasmas they produce in electric discharge. The degree of ionization in such plasmas is usually low, but electron densities of 10^{16} to 10^{18} electrons per cubic metre can be achieved with an electron temperature of 100,000 K. The electrons responsible for current flow are produced by ionization in a region near the cathode, with most of the potential difference between the two electrodes occurring there. This region does not contain a plasma, but the region between it and the anode (i.e., the positive electrode) does.

Other applications of the glow discharge include electronic switching devices; it and similar plasmas produced by radio-frequency techniques can be used to provide ions for particle accelerators and act as generators of laser beams. As the current is increased through a glow discharge, a stage is reached when the energy generated at the cathode is sufficient to provide all the conduction electrons directly from the cathode surface, rather than from gasbetween the electrodes. Under this condition the large cathode potential difference disappears, and the plasma column contracts. This new state of electric discharge is called an arc. Compared with the glow discharge, it is a high-density plasma and will operate over a large range of pressures. Arcs are used as light sources for welding, in electronic switching, for rectification of alternating currents, and in high-temperature chemistry. Running an arc between concentric electrodes and injecting gas into such a region causes a hot, high-density plasma mixture called a plasma jet to be ejected. It has many chemical and metallurgical applications.

Natural Plasmas

Extraterrestrial Forms

It has been suggested that the universe originated as a violent explosion about 13.8 billion years ago and initially consisted of a fireball of completely ionized hydrogen plasma. Irrespective of the truth of this, there is little matter in the universe now that does not exist in the plasma state. The observed stars are composed of plasmas, as are interstellar and interplanetary media and the outer atmospheres of planets. Scientific knowledge of the universe has come primarily from studies of electromagnetic radiation emitted by plasmas and transmitted through them and, since the 1960s, from space probes within the solar system.

In a star the plasma is bound together by gravitational forces, and the enormous energy it emits originates in thermonuclear fusion reactions within the interior. Heat is transferred from the interior to the exterior by radiation in the outer layers, where convection is of greater importance. In the vicinity of a hot star, the interstellar medium consists almost entirely of completely ionized hydrogen, ionized by the star's ultraviolet radiation. Such regions are referred to as H II regions. The greater proportion by far of interstellar medium, however, exists in the form of neutral hydrogen clouds referred to as H I regions. Because the heavy atoms in such clouds are ionized by ultraviolet radiation (or photoionized), they also are considered to be plasmas, although the degree of ionization is probably only one part in 10,000. Other components of the interstellar medium are grains of dust and cosmic rays, the latter consisting of very high-energy atomic nuclei completely stripped of electrons. The almost isotropic velocity distribution of the cosmic rays may stem from interactions with waves of the background plasma.

Throughout this universe of plasma there are magnetic fields. In interstellar space magnetic fields are about 5×10^{-6} gauss (a unit of magnetic field strength) and in interplanetary space 5×10^{-5} gauss, whereas in intergalactic space they could be as low as 10^{-9} gauss. These values are exceedingly small compared with the Earth's surface field of about 5×10^{-1} gauss. Although small in an absolute sense, these fields are nevertheless gigantic, considering the scales involved. For example, to simulate interstellar phenomena in the laboratory, fields of about 10^{15} gauss would be necessary. Thus, these fields play a major role in nearly all astrophysical phenomena. On the Sun the average surface field is in the vicinity of 1 to 2 gauss, but magnetic disturbances arise, such as sunspots, in which fields of between 10 and 1,000 gauss occur. Many other stars are also known to have magnetic fields. Field strengths of 10^{-3} gauss are associated with various extragalactic nebulae from which synchrotron radiation has been observed.

Solar-terrestrial Forms

Regions of the Sun

The visible region of the Sun is the photosphere, with its radiation being about the same as the continuum radiation from a 5,800 K blackbody. Lying above the photosphere is the chromosphere, which is observed by the emission of line radiation from various atoms and ions. Outside the chromosphere, the corona expands into the ever-blowing solar wind, which on passing through the planetary system eventually encounters the interstellar medium. The corona can be seen in spectacular fashion when the Moon eclipses the bright photosphere. During the times in which sunspots are greatest in number (called the sunspot maximum), the corona is very extended and the solar wind is fierce. Sunspot activity waxes and wanes with roughly an 11-year cycle. During the mid-1600s and early 1700s, sunspots virtually disappeared for a period known as the Maunder minimum. This time coincided with the Little Ice Age in Europe, and much conjecture has arisen about the possible effect of sunspots on climate. Periodic variations similar to that of sunspots have been observed in tree rings and lake-bed sedimentation. If real, such an effect is important because it implies that the Earth's climate is fragile.

In 1958 the American astrophysicist Eugene Parker showed that the equations describing the flow of plasma in the Sun's gravitational field had one solution that allowed the gas to become supersonic and to escape the Sun's pull. The solution was much like the description of a rocket nozzle in which the constriction in the flow is analogous to the effect of gravity. Parker predicted the Sun's atmosphere would behave just as this particular solar-wind solution predicts rather than according to the solar-breeze solutions suggested by others. The interplanetary satellite probes of the 1960s proved his solution to be correct.

Interaction of the Solar Wind and the Magnetosphere

The solar wind is a collisionless plasma made up primarily of electrons and protons and carries an outflow of matter moving at supersonic and super-Alfvénic speed. The wind takes with it an extension of the Sun's magnetic field, which is frozen into the highly conducting fluid. In the region of the Earth, the wind has an average speed of 400 kilometres per second; and, when it encounters the planet's magnetic field, a shock front develops, the pressures acting to compress the field on the side toward the Sun and elongate it on the nightside (in the Earth's lee away from the Sun). The Earth's magnetic field is therefore confined to a cavity called the magnetosphere, into which the direct entry of the solar wind is prohibited. This cavity extends for about 10 Earth radii on the Sun's side and about 1,000 Earth radii on the nightside.

Inside this vast magnetic field a region of circulating plasma is driven by the transfer of momentum from the solar wind. Plasma flows parallel to the solar wind on the edges of this region and back toward the Earth in its interior. The resulting system acts as a secondary magneto hydrodynamic generator (the primary one being the solar wind itself). Both generators produce potential on the order of 100,000 volts. The solar-wind potential appears across the polar caps of the Earth, while the magnetospheric potential appears across the auroral oval. The latter is the region of the Earth where energetic electrons and ions precipitate into the planet's atmosphere, creating a spectacular light show. This particle flux is energetic enough to act as a new source of plasmas even when the Sun is no longer shining. The auroral oval becomes a good conductor; and large electric currents flow along it, driven by the potential difference across the system. These currents commonly are on the order of 1,000,000 amperes.

The plasma inside the magnetosphere is extremely hot (1–10 million K) and very tenuous (1–10 particles per cubic centimetre). The particles are heated by a number of interesting plasma effects, the most curious of which is the auroral acceleration process itself. A particle accelerator that may be the prototype for cosmic accelerators throughout the universe is located roughly one Earth radius above the auroral oval and linked to it by all-important magnetic field lines. In this region the auroral electrons are boosted by a potential difference on the order of three to six kilovolts, most likely created by an electric field parallel to the magnetic field lines and directed away from the Earth. Such a field is difficult to explain because magnetic field lines usually act like nearly perfect conductors. The auroras occur on magnetic field lines that—if it were not for the distortion of the Earth's dipole field—would cross the equatorial plane at a distance of 6–10 Earth radii.

Closer to the Earth, within about 4 Earth radii, the planet wrests control of the system away from the solar wind. Inside this region the plasma rotates with the Earth, just as its atmosphere rotates with it. This system can also be thought of as a magnetohydrodynamic generator in which the rotation of the atmosphere and the ionospheric plasma in it create an electric field that puts the inner magnetosphere in rotation about the Earth's axis. Since this inner region is in contact with the dayside of the Earth where the Sun creates copious amounts of plasma in the ionosphere, the inner zone fills up with dense, cool plasma to form the plasmasphere. On a planet such as Jupiter, which has both a larger magnetic field and a higher rotation rate than the Earth, planetary control extends much farther from the surface.

Ionosphere and Upper Atmosphere

At altitudes below about 2,000 kilometres, the plasma is referred to as the ionosphere. Thousands of rocket probes have helped chart the vertical structure of this region of the atmosphere, and numerous satellites have provided latitudinal and longitudinal information. The ionosphere was discovered in the early 1900s when radio waves were found to propagate"over the horizon." If radio waves have frequencies near or below the plasma frequency, they cannot propagate throughout the plasma of the ionosphere and thus do not escape into space; they are instead either reflected or absorbed. At night the absorption is low since little plasma exists at the height of roughly 100 kilometres where absorption is greatest. Thus, the ionosphere acts as an effective mirror, as does the Earth's surface, and waves can be reflected around the entire planet much as in a waveguide. A great communications revolution was initiated by the wireless, which relied on radio waves to transmit audio signals. Development continues to this day with satellite systems that must propagate through the ionospheric plasma. In this case, the wave frequency must be higher than the highest plasma frequency in the ionosphere so that the waves will not be reflected away from the Earth.

The dominant ion in the upper atmosphere is atomic oxygen, while below about 200 kilometres molecular oxygen and nitric oxide are most prevalent. Meteor showers also provide large numbers of metallic atoms of elements such as iron, silicon, and magnesium, which become ionized in sunlight and last for long periods of time. These form vast ion clouds, which are responsible for much of the fading in and out of radio stations at night.

Lower Atmosphere and Surface of the Earth

A more normal type of cloud forms at the base of the Earth's plasma blanket in the summer polar mesosphere regions. Located at an altitude of 85 kilometres, such a cloud is the highest on Earth and can be seen only when darkness has just set in on the planet. Hence, clouds of this kind have been called noctilucent clouds. They are thought to be composed of charged and possibly dusty ice crystals that form in the coldest portion of the atmosphere at a temperature of 120 K. This unusual medium has much in common with dusty plasmas in planetary rings and other cosmic systems. Noctilucent clouds have been increasing in frequency throughout the 20th century and may be a forerunner of global change.

High-energy particles also exist in the magnetosphere. At about 1.5 and 3.5 Earth radii from the centre of the planet, two regions contain high-energy particles. These regions are the Van Allen radiation belts, named after the American scientist James Van Allen, who discovered them using radiation detectors aboard early spacecraft. The charged particles in the belts are trapped in the mirror system formed by the Earth's magnetic dipole field.

Plasma can exist briefly in the lowest regions of the Earth's atmosphere. In a lightning stroke an oxygen-nitrogen plasma is heated at approximately 20,000 K with an ionization of about 20 percent, similar to that of a laboratory arc. Although the stroke is only a few centimetres thick and lasts only a fraction of a second, tremendous energies are dissipated. A lightning flash between the ground and a cloud, on the average, consists of four such strokes in rapid succession. At all times, lightning is occurring somewhere on the Earth, charging the surface negatively with respect to the ionosphere by roughly 200,000 volts, even far from the nearest thunderstorm. If lightning ceased everywhere for even one hour, the Earth would discharge. An associated phenomenon is ball lightning. There are authenticated reports of glowing, floating, stable balls of light several tens of centimetres in diameter occurring at times of intense electrical activity in the atmosphere. On contact with an object, these balls release large amounts of energy. Although lightning balls are probably plasmas, so far no adequate explanation of them has been given.

Considering the origins of plasma physics and the fact that the universe is little more than a vast sea of plasma, it is ironic that the only naturally occurring plasmas at the surface of the Earth besides lightning are those to be found in ordinary matter. The free electrons responsible for electrical conduction in a metal constitute a plasma. Ions are fixed in position at lattice points, and so plasma behaviour in metals is limited to such phenomena as plasma oscillations and electron cyclotron waves (called helicon waves) in which the electron component behaves separately from the ion component. In semiconductors, on the other hand, the current carriers are electrons and positive holes, the latter behaving in the material as free positive charges of finite mass. By proper preparation, the number of electrons and holes can be made approximately equal so that the full range of plasma behaviour can be observed.

Magnetic Fields

The importance of magnetic fields in astrophysical phenomena has already been noted. It is believed that these fields are produced by self-generating dynamos, although the exact details are still not fully understood. In the case of the Earth, differential rotation in its liquid conducting core causes the external magnetic dipole field (manifest as the North and South poles). Cyclonic turbulence in the liquid, generated by heat conduction and Coriolis forces (apparent forces accompanying all rotating systems, including the heavenly bodies), generates the dipole field from these loops. Over geologic time, the Earth's field occasionally becomes small and then changes direction, the North Pole becoming the South Pole and vice versa. During the times in which the magnetic field is small, cosmic rays can more easily reach the Earth's surface and may affect life forms by increasing the rate at which genetic mutations occur.

Similar magnetic-field generation processes are believed to occur in both the Sun and the Milky Way Galaxy. In the Sun the circular internal magnetic field is made observable by lines of force apparently breaking the solar surface to form exposed loops; entry and departure points are what are observed as sunspots. Although the exterior magnetic field of the Earth is that of a dipole, this is further modified by currents in both the ionosphere and magnetosphere. Lunar and solar tides in the ionosphere lead to motions across the Earth's field that produce currents, like a dynamo, that modify the initial field. The auroral oval current systems discussed earlier create even larger magnetic-field fluctuations. The intensity of these currents is modulated by the intensity of the solar wind, which also induces or produces other currents in the magnetosphere. Such currents taken together constitute the essence of a magnetic storm.

PLASMA AS A FLUID

The behavior of most of the plasma effects in ion and Hall thrusters can be described by simplified models in which the plasma is treated as a fluid of neutral particles and electrical charges with Maxwellian distribution functions, and the interactions and motion of only the fluid elements must be considered. Kinetic effects that consider the actual velocity distribution of each species are important in some instances, but will not be addressed here.

Momentum Conservation

In constructing a fluid approach to plasmas, there are three dominant forces on the charged particles in the plasma that transfer momentum that are considered here. First, charged particles react to electric and magnetic field by means of the Lorentz force, which was given by equation,

$$F = m\frac{dv}{dt} = q(E + v \times B).$$

$$F_L = m\frac{dv}{dt} = q(E + v \times B).$$

Next, there is a pressure gradient force,

$$Fp = -\frac{\nabla . p}{n} = -\frac{\nabla(nkT)}{n},$$

Where the pressure is given by $P = nkT$ and should be written more rigorously as a stress tensor since it can, in general, be anisotropic. For plasmas with temperatures that are generally spatially constant, the force due to the pressure gradient is usually written simply as:

$$Fp = -kT\frac{\nabla n}{n}$$

Finally, collisions transfer momentum between the different charged particles, and also between the charged particles and the neutral gas. The force due to collisions is:

$$F_c = -m\sum_{a,b} v_{ab}(v_a - v_b),$$

where v_{ab} is the collision frequency between species a and b. Using these three force terms, the fluid momentum equation for each species is:

$$mn\frac{dv}{dt} = mn\left[\frac{\partial v}{\partial t} + |v.\nabla)v\right] = qn\,(E + V \times B) - \nabla . p - mnv\,(v - v_0),$$

Where the convective derivative has been written explicitly and the collision term must be summed over all collisions. Utilizing conservation of momentum, it is possible to evaluate how the electron fluid behaves in the plasma. For example, in one dimension and in the absence of magnetic fields and collisions with other species, the fluid equation of motion for electrons can be written as:

$$mn_e \left[\frac{\partial v_z}{\partial t} + v\left(v.\nabla \right)v_z \right] = qn_e E_z - \frac{\partial p}{\partial z} \ ,$$

Where v_z is the electron velocity in the z-direction and p represents the electron pressure term. Neglecting the convective derivative, assuming that the velocity is spatially uniform, and using $Fp = -kT\frac{\nabla n}{n}$ gives:

$$m\frac{\partial v_z}{\partial t} = -eE_z - \frac{kT_e}{n_e}\frac{\partial n_e}{\partial_z},$$

Integrating this equation and solving for the electron density produces the Boltzmann relationship for electrons:

$$n_e = n_e\left(0 \right)e^{(e\phi/kT_e)} \ ,$$

where ϕ is the potential relative to the potential at the location of ne (o). Equation is also sometimes known as the barometric law. This relationship simply states that the electrons will respond to electrostatic fields (potential changes) by varying their density to preserve the pressure in the system. This relationship is generally valid for motion along a magnetic field and tends to hold for motion across magnetic fields if the field is weak and the electron collisions are frequent.

Particle Conservation

Conservation of particles and/or charges in the plasma is described by the continuity equation:

$$\frac{\partial n}{\partial t} + \nabla.nv = \dot{n}_s,$$

Where $\dot{n}_s$ represents the time-dependent source or sink term for the species being considered. Continuity equations are sometimes called mass-conservation equations because they account for the sources and sinks of particles into and out of the plasma.

Utilizing continuity equations coupled with momentum conservation and with Maxwell's equations, it is possible to calculate the response rate and wave-like behavior of plasmas. For example, the rate at which a plasma responds to changes in potential is related to the plasma frequency of the electrons. Assume that there is no magnetic field in the plasma or that the electron motion is along the magnetic field in the z-direction. To simplify this derivation, also assume that the ions are fixed uniformly in space on the time scales of interest here due to their large mass, and that there is no thermal motion of the particles (T = o).

Since the ions are fixed in this case, only the electron equation of motion is of interest:

$$mn_e\left[\frac{\partial v_z}{\partial t}+(v.\nabla)v_z\right]=-en_eE_z,$$

and the electron equation of continuity is:

$$\frac{\partial n_e}{\partial t}+\nabla.(n_e\,v)=0.$$

The relationship between the electric field and the charge densities is given by equation which for singly ionized particles can be written using equation. as,

$$\rho=\sum_s q_s n_s=e(Zn_i-n_e),$$

$$\nabla\cdot E=\frac{\rho}{\varepsilon_o}$$

$$\nabla.E_1=\frac{\rho}{\varepsilon_o}=\frac{e}{\varepsilon_o}(n_1-n_e).$$

The wave-like behavior of this system is analyzed by linearization using,

$$E=E\ +E$$
$$v=v_0+v_1$$
$$n=n\ +n$$

where E_0 , v_0 , and no are the equilibrium values of the electric field, electron velocity, and electron density, and E_1 , B_1 , and j_1 are the perturbed values of these quantities. Since quasi-neutral plasma has been assumed, $E_0=0$, and the assumption of a uniform plasma with no temperature means that $n_0=v_0=0$. Likewise, the time derivatives of these equilibrium quantities are zero.

Linearizing equation gives:

$$\nabla.E_1=-\frac{e}{\varepsilon_o}n_1.$$

$$\nabla.E_1=\frac{\rho}{\varepsilon_o}=\frac{e}{\varepsilon_o}(n_1-n_e).$$

Using equation,

$$E=E_0+E_1$$
$$v=v_0+v_1$$

and

$$n = n_o + n_1,$$

in

$$\frac{dv_1}{dt} = -\frac{e}{\varepsilon_o} E_1 \hat{z},$$

equation results in,

$$mn_e \left[\frac{\partial v_z}{\partial t} + (v.\nabla) v_z \right] = -en_e E_z,$$

where the linearized convective derivative has been neglected. Linearizing the continuity equation $\frac{\partial}{\partial} + \nabla.(\quad v) = 0$ gives:

$$\frac{dn_1}{dt} = -n_o \nabla v_1 \hat{z}$$

where the quadratic terms, such as $n_1 v_1$, etc., have been neglected as small. In the linear regime, the oscillating quantities will behave sinusoidally:

$$E_1 = E_1 e^{i(kz-\omega t)} \hat{z}$$
$$v_1 = v_1 e_i^{(kz-\omega t)} \hat{z}$$
$$n_1 = n_1 e^{i(kz-\omega t)}.$$

This means that the time derivates in momentum and continuity equations can be replaced by it and the gradient in equation $\nabla.E_1 = -\frac{e}{\varepsilon_o} n_1$. can be replaced by ik in the

$\hat{z}$ direction. Combining equation $\nabla.E_1 = -\frac{e}{\varepsilon_o} n_1.$, $\frac{dv_1}{dt} = -\frac{e}{\varepsilon_o} E_1 \hat{z}$, and $\frac{dn_1}{dt} = -n_o \nabla v_1 \hat{z}$, using the time and spatial derivatives of the oscillating quantities, and solving for the frequency of the oscillation gives:

$$\omega_p = \left(\frac{n_e e^2}{\varepsilon_o m} \right)^{1/2}$$

where ω_p is the electron plasma frequency. A useful numerical formula for the electron plasma frequency is,

$$f_p = \frac{\omega p}{2\pi} = 9 \sqrt{n_e},$$

Where the plasma density is in m⁻³. This frequency is one of the fundamental parameters of a plasma, and the inverse of this value is approximately the minimum time required for the plasma to react to changes in its boundaries or in the applied potentials. For example, if the plasma density is 1018 m−3, the electron plasma frequency is 9 GHz, and the electron plasma will respond to perturbations in less than a nanosecond.

In a similar manner, if the ion temperature is assumed to be negligible and the gross response of the plasma is dominated by ion motions, the ion plasma frequency can be found to be,

$$\Omega p = \left(\frac{n_e e^2}{\varepsilon_o M} \right)^{1/2}$$

This equation provides the approximate time scale in which ions move in the plasma. For our previous example for a 10^{18} m^{-3} plasma density composed of xenon ions, the ion plasma frequency is about 18 MHz, and the ions will respond to first order in a fraction of a microsecond. However, the ions have inertia and respond at the ion acoustic velocity given by,

$$v_a = \sqrt{\frac{\gamma_i k T_i + k T_e}{M}} \ .$$

Where γ i is the ratio of the ion specific heats and is equal to one for isothermal ions. In the normal case for ion and Hall thrusters, where $T_e \gg T_i$, the ion acoustic velocity is simply,

$$v_a = \sqrt{\frac{k T_e}{M}} \ .$$

It should be noted that if finite-temperature electrons and ions had been included in the derivations above, the electron-plasma and ion-plasma oscillations would have produced waves that propagate with finite wavelengths in the plasma. Electron-plasma waves and ion-plasma waves (sometimes called ion acoustic waves) occur in most electric thruster plasmas with varying amplitudes and effects on the plasma behavior. The dispersion relationships for these waves, which describe the relationship between the frequency and the wavelength of the wave, are derived in detail in plasma textbooks such as Chen and will not be re-derived here.

Energy Conservation

The general form of the energy equation for charged species "s," moving with velocity vs in the presence of species "n" is given by,

$$\frac{\partial}{\partial t} \left(n_s m_s \frac{v_s^2}{2} + \frac{3}{2} p_s \right) + \nabla \left(n_s m_s \frac{v_s^2}{2} + \frac{5}{2} p_s \right) v_s + \nabla \grave{e}_s$$

$$= q_s n_s \, \mathbf{E} + \frac{R_s}{q_s n_s} . v_s + Q_s \text{-} \Psi \, s.$$

For simplicity, equation $\dfrac{\partial}{\partial t}\left(n_s m_s \dfrac{v_s^2}{2}+\dfrac{3}{2}p_s\right)+\nabla\left(n_s m_s \dfrac{v_s^2}{2}+\dfrac{5}{2}p_s\right)v_s+\nabla\grave{e}_s$ neglects

$$= q_s n_s\, \mathbf{E}+\frac{R_s}{q_s n_s}.v_s+ Q_s\text{-}\Psi\ s.$$

viscous heating of the species. The divergence terms on the left-hand side represent the total energy flux, which includes the work done by the pressure, the macroscopic energy flux, and the transport of heat by conduction $\grave{e}s = k_s \Delta T_s$. The thermal conductivity of the species is denoted by k_s, which is given in SI units by:

$$k_s = 3.2\ \frac{\tau e n e^2 T_{eV}}{m},$$

Where TeV in this equation is in electron volts (eV). The right-hand side of equation:

$$\frac{\partial}{\partial t}\left(n_s m_s \frac{v_s^2}{2}+\frac{3}{2}p_s\right)+\nabla\left(n_s m_s \frac{v_s^2}{2}+\frac{5}{2}p_s\right)v_s+\nabla\grave{e}_s$$

$$= q_s n_s\, \mathbf{E}+\frac{R_s}{q_s n_s}.v_s+ Q_s\text{-}\Psi\ s.$$

accounts for the work done by other forces as well for the generation/loss of heat as a result of collisions with other particles. The term R_s represents the mean change in the momentum of particles "s" as a result of collisions with all other particles:

$$R_s\sum_n R_{Sn} = -\sum_n n_s m_s v_{sn}\left(v_s-v_n\right).$$

The heat-exchange terms are Q_s, which is the heat generated/lost in the particles of species "s" as a result of elastic collisions with all other species, and Ψ_s, the energy loss by species "s" as a result of inelastic collision processes such as ionization and excitation.

It is often useful to eliminate the kinetic energy from equation:

$$\frac{\partial}{\partial t}\left(n_s m_s \frac{v_s^2}{2}+\frac{3}{2}p_s\right)+\nabla\left(n_s m_s \frac{v_s^2}{2}+\frac{5}{2}p_s\right)v_s+\nabla\grave{e}_s$$

$$= q_s n_s\, \mathbf{E}+\frac{R_s}{q_s n_s}.v_s+ Q_s\text{-}\Psi\ s.$$

to obtain a more applicable form of the energy conservation law. The left-hand side of equation:

$$\frac{\partial}{\partial t}\left(n_s m_s \frac{v_s^2}{2}+\frac{3}{2}p_s\right)+\nabla\left(n_s m_s \frac{v_s^2}{2}+\frac{5}{2}p_s\right)v_s+\nabla\grave{e}_s$$

$$= q_s n_s\, \mathbf{E}+\frac{R_s}{q_s n_s}.v_s+ Q_s\text{-}\Psi\ s.$$

is expanded as:

$$n_s m_s v_s \frac{Dv_s}{Dt} = \frac{m_s v_s^2}{2} \frac{Dv_s}{Dt} n_s m_s \frac{v_s^2}{2}.\nabla - v_s \frac{3}{2} \frac{\partial n_s}{\partial t} + \nabla . + \left(\frac{5}{2}\right) p_s v_s + \theta_s.$$

$$= q_s n_s E.v_s + R_s.v_s + Q_s - \Psi s.$$

The continuity equation for the charged species is in the form:

$$\frac{Dn_s}{Dt} = \frac{\partial n_s}{\partial t} v_s.\nabla n_s = \dot{n} - n_s \nabla.v_s .$$

Combining these two equations with the momentum equation dotted with V_s gives:

$$n_s m_s v_s \frac{Dv_s}{Dt} = n_s q_s v_s.E - v_s \nabla p_s + v_s R_s - \dot{n} m_s v_s^2.$$

The energy equation can now be written as:

$$\frac{3}{2} \frac{\partial p_s}{\partial t} + \nabla \left(\frac{5}{2} p_s v_s + \theta_s\right) - v_s.\nabla ps = Q_s - \Psi_s - \dot{n} \frac{m_s v_s^2}{2}.$$

The heat-exchange terms for each species Q_s consists of "frictional" (denoted by super-script R) and "thermal" (denoted by superscript T) contributions:

$$Qs = Q_S^R + Q_S^T ,$$

$$Q_S^R \equiv \sum_n R_{sn} .v_s,$$

$$Q_e^T \equiv -\sum_n n_S \frac{2m_s}{m_a} v_{sn} \frac{3}{2}\left(\frac{kT_s}{e} - \frac{kT_n}{e}\right)$$

In a partially ionized gas consisting of electrons, singly charged ions, and neutrals of the same species, the frictional and thermal terms for the electrons take the form:

$$Q_S^R = \left(\frac{R_{ei} + R_{en}}{en_e}\right).J_e = \left(E + \frac{\nabla p_e}{en_e}\right). J_e.$$

$$Q_e^T = -\left[3n_e \frac{m}{M} v_{ei} \frac{k}{e}(T_e - T_i) v_{en} \frac{k}{e}(T_e - T_i)\right]$$

where as usual M denotes the mass of the heavy species, and the temperature of the ions and neutrals is denoted by T_i and T_n , respectively. Using the steadystate electron momentum equation, in the absence of electron inertia, it is possible to write:

$$Q_S^R = \left(\frac{R_{ei} + R_{en}}{en_e}\right).J_e = \left(E + \frac{\nabla p_e}{en_e}\right). J_e.$$

Thus equation $\frac{3}{2}\frac{\partial p_s}{\partial t} + \nabla\left(\frac{5}{2}\, p_s\, \mathrm{v}_s + \theta_s\right) - v_s.\nabla ps = Q_s - \Psi_s - \dot{n}\frac{m_s v_s^{\,2}}{2}$. for the electrons becomes:

$$\frac{3}{2}\frac{\partial p_e}{\partial t} + \nabla\left(\frac{5}{2}\, p_e\, v_e + \theta_e\right) - Qe - J_e \cdot \frac{\nabla p_e}{en}\,\dot{n}eU_i$$
$$= E.\,J_e - \dot{n}eU_i,$$

where the inelastic term is expressed as $\Psi_e = e\dot{n}U_i$ to represent the electron energy loss due to ionization, with U_i (in volts) representing the first ionization potential of the atom. In equation:

$$\frac{3}{2}\frac{\partial p_e}{\partial t} + \nabla\left(\frac{5}{2}\, p_e\, v_e + \theta_e\right) - Qe - J_e \cdot \frac{\nabla p_e}{en}\,\dot{n}eU_i$$
$$= E.\,J_e - \dot{n}eU_i,$$

,

the $m_e v_e^2 / 2$ correction term has been neglected because usually in ion and Hall thrusters $eU_i \gg m_e v_e^2 / 2$. If multiple ionization and excitation losses are significant, the inelastic terms in equation:

$$\frac{3}{2}\frac{\partial p_{in}}{\partial t} + \nabla\left(\frac{5}{2}\, p_e\, v_e + \theta_e\right) = Qe - J_e \cdot \frac{\nabla p_e}{en} - \dot{n}eU_i,$$
$$= E.\,J_e - \dot{n}eU_i,$$

must be augmented accordingly.

In ion and Hall thrusters, it is common to assume a single temperature or distribution of temperatures for the heavy species without directly solving the energy equations. In some cases, however, such as in the plume of a hollow cathode for example, the ratio of T_e/T_i is important for determining the extent of Landau damping on possible electrostatic instabilities. The heavy species temperature is also important for determining the total pressure inside the cathode. Thus, separate energy equations must be solved directly. Assuming that the heavy species are slow moving and the inelastic loss terms are negligible, equation:

$$\frac{3}{2}\frac{\partial p_s}{\partial t} + \nabla\left(\frac{5}{2}\, p_s\, \mathrm{v}_s + \theta_s\right) - Q_s = \Psi_s - \dot{n}\frac{m_s v_s^{\,2}}{2}.$$ for ions takes the form,

$$\frac{3}{2}\frac{\partial p_{in}}{\partial t} + \nabla\left(\frac{5}{2}\, p_{in}\, \mathrm{v}_{in} + \theta in\right) - \mathrm{v}_{in}.\nabla p_{in} = Q_{in},$$

where the subscript "in" represents ion-neutral collisions. Finally, the total heat generated in partially ionized plasmas as a result of the (elastic) friction between the various species is given by:

$$\sum_s Q_S^R = Q_e^R + Q_i^R + Q_n^R$$

$$= -\left(R_{ei} + R_{en} \right).v_e - \left(R_{ie} + R_{in} \right).v_i - \left(R_{ne} + R_{ni} \right).v_n.$$

Since $R_{sa} = -R_{as}$, it is possible to write this as:

$$\sum_s Q_S^R = -R_{ei}\left(v_e - v_i \right) R_{en}\left(v_e - v_n \right) - R_{in}.\left(v_i - v_n \right).$$

The energy conservation equations can be used with the momentum and continuity equations to provide a closed set of equations for analysis of plasma dynamics within the fluid approximations.

Sheaths at the Boundaries of Plasmas

While the motion of the various particles in the plasma is important in understanding the behavior and performance of ion and Hall thrusters, the boundaries of the plasma represent the physical interface through which energy and particles enter and leave the plasma and the thruster. Depending on the conditions, the plasma will establish potential and density variations at the boundaries in order to satisfy particle balance or the imposed electrical conditions at the thruster walls. This region of potential and density change is called the sheath, and understanding sheath formation and behavior is also very important in understanding and modeling ion and Hall thruster plasmas.

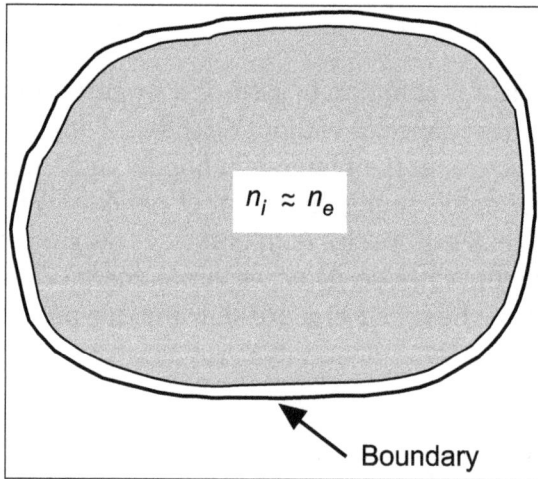

$$n_i \approx n_e$$

Boundary

Generic quasi-neutral plasmaenclosed in a boundary.

Consider the generic plasma in figure, consisting of quasi-neutral ion and electron densities with temperatures given by T_i and T_e , respectively. The ion current density to the boundary "wall" for singly charged ions, to first order, is given by n_iev_i , where vi is the ion velocity. Likewise, the electron flux to the boundary wall, to first order, is given by

$n_e e v_e$, where v_e is the electron velocity. The ratio of the electron flux to the ion current density going to the boundary, assuming quasi-neutrality, is:

$$\frac{J_e}{J_i} = \frac{n_e e v_e}{n_i e v_i} = \frac{v_e}{v_i}.$$

In the absence of an electric field in the plasma volume, conservation of energy for the electrons and ions is given by:

$$\frac{1}{2} m v_e^2 = \frac{kT_e}{e},$$

$$\frac{1}{2} m v_i^2 = \frac{kT_i}{e},$$

If it is assumed that the electrons and ions have the same temperature, the ratio of current densities to the boundary is:

$$\frac{J_e}{J_i} = \frac{v_e}{v_i} = \sqrt{\frac{M}{m}}.$$

Table shows the mass ratio M/m for several gas species. It is clear that the electron current out of the plasma to the boundary under these conditions is orders of magnitude higher than the ion current due to the much higher electron mobility. This would make it impossible to maintain the assumption of quasineutrality in the plasma used in equation $\frac{J_e}{J_i} = \frac{n_e e v_e}{n_i e v_i} = \frac{v_e}{v_i}$. because the electrons would leave the volume much faster than the ions.

If different temperatures between the ions and electrons are allowed, the ratio of the current densities to the boundary is:

$$\frac{J_e}{J_i} = \frac{v_e}{v_i} = \sqrt{\frac{M}{m} \frac{T_e}{T_i}}.$$

To balance the fluxes to the wall to satisfy charge continuity (an ionization event makes one ion and one electron), the ion temperature would have to again be orders of magnitude higher than the electron temperatures. In ion and Hall thrusters, the opposite is true and the electron temperature is normally about an order of magnitude higher than the ion temperature, which compounds the problem of maintaining quasi-neutrality in a plasma.

Table: Ion-to-electron mass ratios for several gas species.

Gas	Mass ratio M/m	Square root of the mass ratio M/m
Protons (H+)	1836	42.8

Argon	73440	270.9
Xenon	241066.8	490.8

In reality, if the electrons left the plasma volume faster than the ions, a charge imbalance would result due to the large net ion charge left behind. This would produce a positive potential in the plasma, which creates a retarding electric field for the electrons. The electrons would then be slowed down and retained in the plasma. Potential gradients in the plasma and at the plasma boundary are a natural consequence of the different temperatures and mobilities of the ions and electrons. Potential gradients will develop at the wall or next to electrodes inserted into the plasma to maintain quasi-neutrality between the charged species. These regions with potential gradients are called sheaths.

Debye Sheaths

To start an analysis of sheaths, assume that the positive and negative charges in the plasma are fixed in space, but have any arbitrary distribution. It is then possible to solve for the potential distribution everywhere using Maxwell's equations. The integral form of equation $\nabla \cdot E = \dfrac{\rho}{\varepsilon_o}$ is Gauss's law:

$$\oint_s E \cdot = ds = \frac{1}{\varepsilon_o} \int_v \rho dV = \frac{Q}{\varepsilon_o},$$

where Q is the total enclosed charge in the volume V and s is the surface enclosing that charge. If an arbitrary sphere of radius r is drawn around the enclosed charge, the electric field found from integrating over the sphere is:

$$E = \frac{Q}{4\pi\varepsilon_o r^2}\hat{r}.$$

Since the electric field is minus the gradient of the potential, the integral form of equation $\rho = \sum_s q_s n_s = e(Zn_i - n_e)$, can be written:

$$\phi 2 - \phi 1 = -\int_{p^1}^{p^2} E \cdot d1,$$

where the integration proceeds along the path $d1$ from point p1 to point p2. Substituting equation $E = \dfrac{Q}{4\pi\varepsilon_o r^2}\hat{r}$. into equation $\phi 2 - \phi 1 = -\int_{p^1}^{} E \cdot d1$, and integrating gives:

$$\phi = \frac{Q}{4\pi\varepsilon_o r}$$

The potential decreases as 1/r moving away from the charge.

However, if the plasma is allowed to react to a test charge placed in the plasma, the potential has a different behavior than predicted by eq. $\phi = \dfrac{Q}{4\pi\varepsilon_o r}$. Utilizing equation $E = -\nabla\phi$ for the electric field in equation $\nabla\cdot E = \dfrac{\rho}{\varepsilon_o}$ gives Poisson's equation:

$$\nabla^2\phi = -\frac{\rho}{\varepsilon_o} = -\frac{e}{\varepsilon_o}\left(Zn_i - n_e\right),$$

Where the charge density in Eq. $\rho = \sum_s q_s n_s = e\left(Zn_i - n_e\right)$, has been used. Assume that the ions are singly charged and that the potential change around the test charge is small $(e\phi \ll kT_e)$, such that the ion density is fixed and $n_i = n_o$. Writing Poisson's equation in spherical coordinates and using equation $n_e = n_e(0)e^{(e\phi/kT_e)}$, to describe the Boltzmann electron density behavior gives:

$$\frac{1}{r^2}\frac{\partial}{\partial r}\left(r^2\frac{\partial\phi}{\partial r}\right)\frac{en_o}{\varepsilon_o}\left[\frac{e\phi}{kT_e} + \frac{1}{2}\left(\frac{e\phi}{kT_e}\right)^2 + ..\right].$$

Neglecting all the higher-order terms, the solution of equation above,

$$\frac{1}{r^2}\frac{\partial}{\partial r}\left(r^2\frac{\partial\phi}{\partial r}\right)\frac{en_o}{\varepsilon_o}\left[\frac{e\phi}{kT_e} + \frac{1}{2}\left(\frac{e\phi}{kT_e}\right)^2 + ..\right].$$

can be written:

$$\phi = \frac{e}{4\pi\varepsilon_o r}\exp\left(-r/\sqrt{\frac{\varepsilon_o kT_e}{n_o e^2}}\right).$$

By defining,

$$\lambda_D = \sqrt{\frac{\varepsilon_o kT_e}{n_o e^2}}$$

as the characteristic Debye length, eq. $\lambda_D = \sqrt{\dfrac{\varepsilon_o kT_e}{n_o e^2}}$ can be written:

$$\phi = \frac{e}{4\pi\varepsilon_o r}\exp\left(-\frac{r}{\lambda_D}\right).$$

This equation shows that the potential would normally fall off away from the test charge inserted in the plasma as 1/r, as previously found, except that the electrons in the plasma have reacted to shield the test charge and cause the potential to decrease exponentially away from it. This behavior of the potential in the plasma is, of course, true for any structure such as a grid or probe that is placed in the plasma and that has a net charge on it.

The Debye length is the characteristic distance over which the potential changes for potentials that are small compared to kT_e. It is common to assume that the sheath around an object will have a thickness of the order of a few Debye lengths in order for the potential to fall to a negligible value away from the object. As an example, consider a plasma with a density of 10^{17} m^{-3} and an electron temperature of 1 eV. Boltzmann's constant k is 1.3807×10^{-23} J/K and the charge is 1.6022×10^{-19} coulombs, so the temperature corresponding to 1 electron volt is,

$$T = 1 \left(\frac{e}{k} \right) = \frac{1.6022 \times 10^{-19}}{1.3807 \times 10^{-23}} = 11604.3 \text{ K}.$$

The Debye length, using the permittivity or free space as 8.85×10^{-12} F/m is then,

$$\lambda_D = \left[\frac{\left(8.85 \times 10^{-12}\right)\left(1.38 \times 10^{-23}\right)11604}{10^{17}\left(1.6 \times 10^{-19}\right)^2} \right]^{1/2}$$

$$= 2.35 \times 10 - 5\,\text{m} = 23.5\,\mu\text{m}$$

A simplifying step to note in this calculation is that kT_e / e in eq. $\lambda_D = \sqrt{\dfrac{\varepsilon_o kT_e}{n_o e^2}}$ has units

of electron volts. A handy formula for the Debye length is $_D(cm)$ $740\sqrt{Tev/n_o}$, where T_{ev} is in electron volts and n_o is in cm^{-3}.

Pre-sheaths

In the previous section, the sheath characteristics for the case of the potential difference between the plasma and an electrode or boundary being small compared to the electron temperature $\left(e\phi \ll kT_e\right)$, was analyzed and resulted in Debye shielding sheaths. What happens for the case of potential differences on the order of the electron temperature? Consider a plasma in contact with a boundary wall, as illustrated in figure. Assume that the plasma is at a reference potential Φ at the center (which can be arbitrarily set), and that cold ions fall through an arbitrary potential of ϕ_o as they move toward the boundary. Conservation of energy states that the ions arrived at the sheath edge with an energy given by,

$$\frac{1}{2} Mv_o^2 = e\phi_o$$

This potential drop between the center of the plasma and the sheath edge, ϕ_o, is called the pre-sheath potential. Once past the sheath edge, the ions then gain an additional energy given by,

$$\frac{1}{2} Mv^2 = \frac{1}{2} Mv_0^2 - e\phi(x),$$

Where v is the ion velocity in the sheath and ϕ is the potential in the sheath (becoming more negative relative to the center of the plasma). Using eq. $\frac{1}{2}Mv_o^2 = e\phi o$ in eq. $\frac{1}{2}Mv^2 = \frac{1}{2}Mv_0^2 - e\phi(x)$ and solving for the ion velocity in the sheath gives.

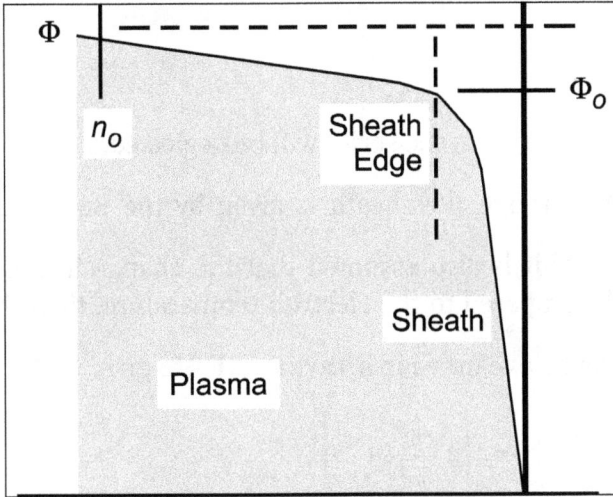

Plasma in contact with a boundary.

$$v = \sqrt{\frac{2e}{M}}\left[\phi_o - \phi\right]^{1/2}.$$

However, from eq. $\frac{1}{2}Mv_o^2 = e\phi o$, $v_o = \sqrt{2e\phi o./M}$, so eq. $v = \sqrt{\frac{2e}{M}}\left[\phi_o - \phi\right]^{1/2}$. can be rearranged to give:

$$\frac{v_o}{v} = \sqrt{\frac{\phi_o}{\phi_o - \phi}}$$

Which represents an acceleration of the ions toward the wall. The ion flux during this acceleration is conserved:

$$n_i v = n_o v_o$$

$$n_i = n_o \frac{v_o}{v}.$$

Using eq. $\frac{v_o}{v} = \sqrt{\frac{\phi_o}{\phi_o - \phi}}$ in eq. $\begin{array}{c} n_i v = n_o v_o \\ n_i = n_o \dfrac{v_o}{v} \end{array}$, the ion density in the sheath is,

$$n_i = n_o \sqrt{\frac{\phi_o}{\phi_o - \phi}}$$

Examining the potential structure close to the sheath edge such that is small compared to the pre-sheath potential o , eq. $n_i = n_o \sqrt{\dfrac{\phi_o}{\phi_o - \phi}}$ can be expanded in a Taylor series to give:

$$n_i = n_o \left(1 - \frac{1}{2} \frac{\phi}{\phi_o} + ... \right),$$

where the higher-order terms in the series will be neglected.

The electron density through the sheath is given by the Boltzmann relationship in eq. $n_e = n_e(0) e^{(e\phi/kT_e)}$. If it is also assumed that the change in potential right at the sheath edge is small compared to the electron temperature, then the exponent in eq. $n_e = n_e(0) e^{(e\phi/kT_e)}$, can be expanded in a Taylor series to give:

$$n_e = n_o \exp\left(\frac{e\phi}{kTe} \right) = n_o \left[1 - \frac{e\phi}{kT_e} + ... \right].$$

Using Eqs. $n_i = n_o \left(1 - \frac{1}{2} \frac{\phi}{\phi_o} + ... \right)$, and $n_e = n_o \exp\left(\frac{e\phi}{kTe} \right) = n_o \left[1 - \frac{e\phi}{kT_e} + ... \right]$. in Poisson's equation, eq. $m \dfrac{\partial v_z}{\partial t} = -eE_z - \dfrac{kT_e}{n_e} \dfrac{\partial n_e}{\partial z}$, for singly charged ions in one dimension gives:

$$\frac{d^2\phi}{dx^2} = -\frac{e}{\varepsilon_o}(n_i - n_e) = -\frac{en_o}{\varepsilon_o}\left[1 - \frac{1}{2}\frac{\phi}{\phi_o} - 1 + \frac{e\phi}{kTe} \right]$$

$$= \frac{en_o\phi}{\varepsilon_o}\left[\frac{1}{2\phi_o} - \frac{e}{kT_e} \right]$$

In order to avoid a positive-going inflection in the potential at the sheath edge, which would then slow or even reflect the ions going into the sheath, the righthand side of eq.

$$\frac{d^2\phi}{dx^2} = -\frac{e}{\varepsilon_o}(n_i - n_e) = -\frac{en_o}{\varepsilon_o}\left[1 - \frac{1}{2}\frac{\phi}{\phi_o} - 1 + \frac{e\phi}{kTe} \right]$$

$$= \frac{en_o\phi}{\varepsilon_o}\left[\frac{1}{2\phi_o} - \frac{e}{kT_e} \right]$$

must always be positive, which implies,

$$\frac{1}{2\phi_o} > \frac{e}{kT_e}.$$

This expression can be rewritten as:

$$\phi_o > \frac{kT_e}{2e}.$$

Which is the Bohm sheath criterion that states that the ions must fall through a potential in the plasma of at least $T_e/2$ before entering the sheath to produce a monotonically decreasing sheath potential. Since $v_o = \sqrt{2e\phi o./M}$, Eq. $\phi_o > \frac{kT_e}{2e}$. can be expressed in familiar form as:

$$v_o > \frac{kT_e}{M}.$$

This is usually called the Bohm velocity for ions entering a sheath. States that the ions must enter the sheath with a velocity of at least $\sqrt{kT_e/M}$ (known as the acoustic velocity for cold ions) in order to have a stable (monotonic) sheath potential behavior. The plasma produces a potential drop of at least $T_e/2$ prior to the sheath (in the pre-sheath region) in order to produce this ion velocity. While not derived here, if the ions have a temperature T_i , it is easy to show that the Bohm velocity will still take the form of the ion acoustic velocity given by,

$$v_o = \sqrt{\frac{\gamma_i kT_i + kT_e}{M}}$$

It is important to realize that the plasma density decreases in the pre-sheath due to ion acceleration toward the wall. This is easily observed from the Boltzmann behavior of the plasma density. In this case, the potential at the sheath edge has fallen to a value of $-kT_e/2e$ relative to the plasma potential where the density is n_o (far from the edge of the plasma). The electron density at the sheath edge is then,

$$n_e = n_o \exp\left(\frac{e\phi}{kTe}\right) = n_o \exp\left[\left(\frac{e\phi}{kTe}\right)\left(\frac{-kTe}{2e}\right)\right]$$

$$= 0.6606\, n_o.$$

Therefore, the plasma density at the sheath edge is about 60% of the plasma density in the center of the plasma.

The current density of ions entering the sheath at the edge of the plasma can be found from the density at the sheath edge in equation:

$$n_e = n_o \exp\left(\frac{e\phi}{kTe}\right) = n_o \exp\left[\left(\frac{e\phi}{kTe}\right)\left(\frac{-kTe}{2e}\right)\right]$$

$$= 0.6606\, n_o.$$

and the ion velocity at the sheath edge in eq.

$$v_o = \sqrt{\frac{\gamma_i kT_i + kT_e}{M}}$$

$$J_i = 0.6 n_o \, ev_i \approx \frac{1}{2} ne \sqrt{\frac{kTe}{M}},$$

Where n is the plasma density at the start of the pre-sheath, which is normally considered to be the center of a collisionless plasma or one collision-mean-free path from the sheath edge for collisional plasmas. It is common to write eq. $J_i = 0.6 n_o \, ev_i \approx \frac{1}{2} ne \sqrt{\frac{kTe}{M}}$, as:

$$I_i \approx \frac{1}{2} ne \sqrt{\frac{kTe}{M}} \, A,$$

Where A is the ion collection area at the sheath boundary. This current is called the Bohm current. For example, consider a xenon ion thruster with a 10^{18} m^{-3} plasma density and an electron temperature of 3 eV. The current density of ions to the boundary of the ion acceleration structure is found to be 118 A/m^2, and the Bohm current to an area of 10^{-2} m^2 is 1.18 A.

Child–langmuir Sheaths

The simplest case of a sheath in a plasma is obtained when the potential across the sheath is sufficiently large that the electrons are repelled over the majority of the sheath thickness. This will occur if the potential is very large compared to the electron temperature ($\phi \gg kT_e / e$). This means that the electron density goes to essentially zero relatively close to the sheath edge, and the electron space charge does not significantly affect the sheath thickness. The ion velocity through the sheath is given by Eq.

$v = \sqrt{\frac{2e}{M}} \left[\phi_o - \phi \right]^{1/2}$. The ion current density is then,

$$J_i = n_i ev = \sqrt{\frac{2e}{M}} [\phi_o - \phi]^{1/2}$$

Solving Eq. $J_i = n_i ev = \sqrt{\frac{2e}{M}} [\phi_o - \phi]^{1/2}$ for the ion density, Poisson's equation in one dimension and with the electron density contribution neglected is,

$$\frac{d^2 \phi}{dx^2} = -\frac{en_i}{\varepsilon_o} = -\frac{J_i}{\varepsilon_o} \left(\frac{M}{2e(\phi_o - \phi)} \right)^{1/2}.$$

The first integral can be performed by multiplying both sides of this equation by $d\phi / dx$ and integrating to obtain,

$$\frac{1}{2}\left[\left(\frac{d\phi}{dx}\right)^2 - \left(\frac{d\phi}{dx}\right)^2_{x=o}\right] = \frac{2J_i}{\varepsilon_o}\left[\frac{M(\phi_o - \phi)}{2e}\right]^{1/2}$$

Assuming that the electric field ($d\phi / dx$) is negligible at x = o, eq.

$$\frac{1}{2}\left[\left(\frac{d\phi}{dx}\right)^2 - \left(\frac{d\phi}{dx}\right)^2_{x=o}\right] = \frac{2J_i}{\varepsilon_o}\left[\frac{M(\phi_o - \phi)}{2e}\right]^{1/2}$$

becomes,

$$\frac{d\phi}{dx} = 2\left(\frac{J_i}{\varepsilon_o}\right)^{1/2}\left[\frac{M(\phi_o - \phi)}{2e}\right]^{1/4}.$$

Integrating this equation and writing the potential across the sheath of thickness d as the voltage V gives the familiar form of the Child–Langmuir law:

$$J_i = \frac{4\varepsilon_o}{9}\left(\frac{2e}{M}\right)^{1/2}\frac{V^{3/2}}{d^2}.$$

This equation was originally derived by Child in 1911 and independently derived by Langmuir in 1913. Equation $J_i = \frac{4\varepsilon_o}{9}\left(\frac{2e}{M}\right)^{1/2}\frac{V^{3/2}}{d^2}$. states that the current per unit area that can pass through a planar sheath is limited by space-charge effects and is proportional to the voltage to the 3/2 power divided by the sheath thickness squared. In ion thrusters, the accelerator structure can be designed to first order using the Child–Langmuir equation where d is the gap between the accelerator electrodes. The Child–Langmuir equation can be conveniently written as:

$$J_e = 2.33\times10^{-6}\frac{V^{3/2}}{d^2}\ \text{electrons}$$

$$J_i = \frac{5.45\times10^{-8}}{\sqrt{M_a}}\frac{V^{3/2}}{d^2}\ \text{singly charged ions}$$

$$= 4.75\times10^{-9}\frac{V^{3/2}}{d^2}\ \text{xenon ions}$$

where M_a is the ion mass in atomic mass units. For example, the spacecharge- limited xenon ion current density across a planar 1-mm grid gap with 1000 V applied is 15 mA/cm².

Generalized Sheath Solution

To find the characteristics of any sheath without the simplifying assumptions used in the above sections, the complete solution to Poisson's equation at a boundary must be obtained. The ion density through a planar sheath, from Eq. $n_i = n_o \sqrt{\dfrac{\phi_o}{\phi_o - \phi}}$, can be written as:

$$n_i = n_o \left(1 - \frac{\phi}{\phi_o} \right)^{-1/2},$$

and the electron density is given by eq:

$$n_e = n_o \, \exp\left(\frac{e\phi}{kT_e} \right).$$

Poisson's equation,

$$\frac{d^2\phi}{dx^2} = -\frac{e}{\varepsilon_o} = n_i - n_e = -\frac{en_o}{\varepsilon_o} \left[\left(1 - \frac{\phi}{\phi_o} \right)^{-1/2} - \exp\left(\frac{e\phi}{kT_e} \right) \right].$$

for singly charged ions then becomes,

$$\frac{d^2\phi}{dx^2} = -\frac{e}{\varepsilon_o} = n_i - n_e = -\frac{en_o}{\varepsilon_o} \left[\left(1 - \frac{\phi}{\phi_o} \right)^{-1/2} - \exp\left(\frac{e\phi}{kT_e} \right) \right].$$

Defining the following dimensionless variables,

$$\chi = -\frac{e\phi}{kT_e},$$

$$\chi_o = \frac{e\phi_o}{kT_e},$$

$$\xi = \frac{x}{\lambda_D},$$

Poisson's equation becomes,

$$\frac{d^2\chi}{d\xi^2} = \left(1 + \frac{\chi}{\chi_o} \right)^{-1/2} - e^{-\chi}.$$

This equation can be integrated once by multiplying both sides by the first derivative of χ and integrating from $\xi_1 = 0$ to $\xi_1 = \xi$:

$$\int_0^\xi \frac{\partial \chi}{\partial \xi} \frac{\partial^2 \chi}{\partial \xi^2} \, d\xi_1 = \int_0^\xi \left(1 + \frac{\chi}{\chi_o}\right)^{-1/2} \partial \chi - \int_0^\xi e^{-\chi} d\chi.$$

where ξ_1 is a dummy variable. The solution to eq.

$$\int_0^\xi \frac{\partial \chi}{\partial \xi} \frac{\partial^2 \chi}{\partial \xi^2} \, d\xi_1 = \int_0^\xi \left(1 + \frac{\chi}{\chi_o}\right)^{-1/2} \partial \chi - \int_0^\xi e^{-\chi} d\chi.$$

is

$$\int_0^\xi \frac{\partial \chi}{\partial \xi} \frac{\partial^2 \chi}{\partial \xi^2} \, d\xi_1 = \int_0^\xi \left(1 + \frac{\chi}{\chi_o}\right)^{-1/2} \partial \chi - \int_0^\xi e^{-\chi} d\chi.$$

Since the electric field $(d\phi / dx)$ is zero away from the sheath where $\xi = 0$, rearrangement of eq.

$$\frac{1}{2}\left[\left(\frac{\partial \chi}{\partial \xi}\right)^2 - \left(\frac{\partial \chi}{\partial \xi}\right)^2_{\xi=0}\right] = 2\chi_o\left[\left(1 + \frac{\chi}{\chi_o}\right)^{1/2} - 1\right] + e^{-\chi} - 1. \text{yields}$$

$$\frac{\partial \chi}{\partial \xi} = \left[4\chi_o\left(1 + \frac{\chi}{\chi_o}\right)^{1/2} = 2e^{-\chi} - 2(2\chi_o - 1)\right]^{1/2}$$

To obtain a solution for $\chi(\xi)$, eq.

$$\frac{\partial \chi}{\partial \xi} = \left[4\chi_o\left(1 + \frac{\chi}{\chi_o}\right)^{1/2} = 2e^{-\chi} - 2(2\chi_o - 1)\right]^{1/2}$$

must be solved numerically. However, as was shown earlier for eq.

$$\frac{d^2\phi}{dx^2} = -\frac{e}{\varepsilon_o}(n_i - n_e) = -\frac{en_o}{\varepsilon_o}\left[1 - \frac{1}{2}\frac{\phi}{\phi_o} - 1 + \frac{e\phi}{kTe}\right]$$

$$= \frac{en_o\phi}{\varepsilon_o}\left[\frac{1}{2\phi_o} - \frac{e}{kT_e}\right]$$

,

the right-hand side must always be positive or the potential will have an inflection at or near the sheath edge. Expanding the right-hand side in a Taylor series and neglecting the higher-order terms, this equation will also produce the Bohm sheath criterion and specify that the ion velocity at the sheath edge must equal or exceed the ion acoustic (or Bohm) velocity. An examination of eq. $\dfrac{\partial \chi}{\partial \xi} = \left[4\chi_o \left(1 + \dfrac{\chi}{\chi_o} \right)^{1/2} = 2e^{-\chi} - 2(2\chi_o - 1) \right]^{1/2}$

shows that the Bohm sheath criterion forces the ion density to always be larger than the electron density through the pre-sheath and sheath, which results in the physically realistic monotonically decreasing potential behavior through the sheath.

Figure shows a plot of the sheath thickness d normalized to the Debye length versus the potential drop in the sheath normalized to the electron temperature. The criterion for a Debye sheath, the potential drop be much less than the electron temperature $(e\phi \ll kT_e)$ which is on the far left-hand side of the graph. The criterion for a Child–Langmuir sheath, the sheath potential be large compared to the electron temperature $(e\phi \gg kT_e)$, which occurs on the righthand side of the graph. This graph illustrates the rule-of-thumb that the sheath thickness is several Debye lengths until the full Child–Langmuir conditions are established. Beyond this point, the sheath thickness varies as the potential to the 3/2 power for a given plasma density.

The reason for examining this general case is because sheaths with potential drops on the order of the electron temperature or higher are typically found at both the anode and insulating surfaces in ion and Hall thrusters. For example, it will be shown later that an insulating surface exposed to a xenon plasma will self-bias to a potential of about $6T_e$, which is called the *floating potential*. For a plasma with an electron temperature of 4 eV and a density of $10^{18}\mathrm{m}^{-3}$, the Debye length from eq. $\lambda_D = \sqrt{\dfrac{\varepsilon_o kT_e}{n_o e^2}}$ is

$1.5 \times 10{-}5\mathrm{m}$. Since the potential is actually significantly greater than the electron temperature, the sheath thickness is several times this value and the sheath transitions to a Child–Langmuir sheath.

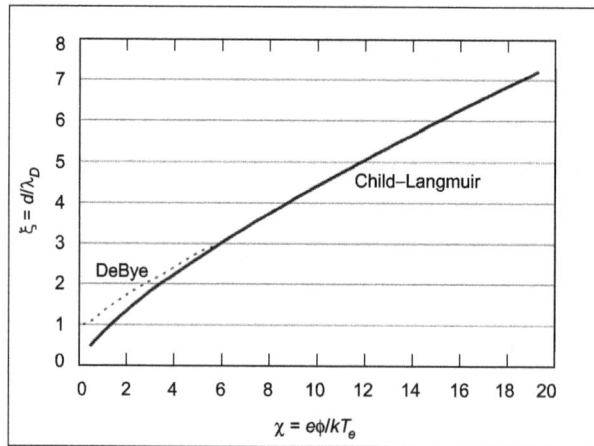

Normalized sheath thickness as a function of the normalized sheath potential showing the transition to a Child–Langmuir sheath as the potential becomes large compared to the electron temperature.

Double Sheaths

So far, only plasma boundaries where particles from the plasma are flowing toward a wall have been considered. At other locations in ion and Hall thrusters, such as in some cathode and accelerator structures, a situation may exist where two plasmas are in contact but at different potentials, and ion and electron currents flow between the plasmas in opposite directions. This situation is called a double sheath, or double layer, and is illustrated in figure. In this case, electrons flow from the zero-potential boundary on the left, and ions flow from the boundary at a potential ϕ_s s on the right. Since the particle velocities are relatively slow near the plasma boundaries before the sheath acceleration takes place, the local space-charge effects are significant and the local electric field is reduced at both boundaries. The gradient of the potential inside the double layer is therefore much higher than in the vacuum case where the potential varies linearly in between the boundaries.

The boundary on the left is at zero potential and that the particles arrive at the sheath edge on both sides of the double layer with zero initial velocity. The potential difference between the surfaces accelerates the particles in the opposite direction across the double layer. The electron conservation of energy gives:

$$\frac{1}{2}mv_e^2 = e\phi$$

$$v_e = \left(\frac{2e\phi}{m}\right)^{1/2}$$

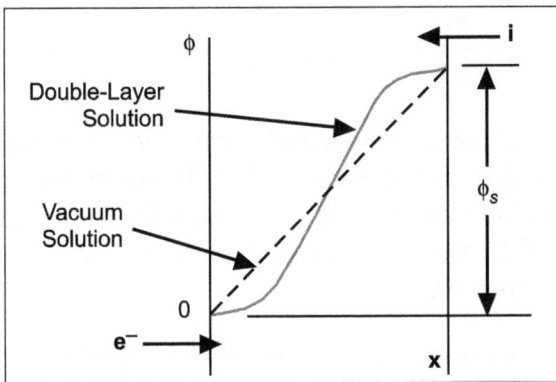

Schematic of the double-layer potential distribution.

and the ion energy conservation gives:

$$\frac{1}{2}Mv_i^2 = e(\phi_s - \phi)$$

$$v_i \left[\frac{2e}{M}(\phi_s - \phi)\right]^{1/2}$$

The charge density in eq. $\rho = \sum_s q_s n_s = e(Zn_i - n_e)$, can be written,

$$\rho = \rho_i + \rho_e$$

$$= \frac{J_i}{v_i} - \frac{J_e}{v_e} = \frac{J_i}{\sqrt{\phi_s - \phi}}\sqrt{\frac{M}{2e}} - \frac{J_e}{\sqrt{\phi}}\sqrt{\frac{m}{2e}}.$$

Poisson's equation can then be written in one dimension as:

$$\frac{dE}{dx} = \frac{\rho}{\varepsilon_o} = \frac{J_i}{\varepsilon_o\sqrt{\phi_s - \phi}}\sqrt{\frac{M}{2e}} - \frac{J_e}{\varepsilon_o\sqrt{\phi}}\sqrt{\frac{m}{2e}}.$$

Integrating once gives:

$$\frac{\varepsilon_o}{2}E^2 = 2J_i\sqrt{\frac{M}{2e}}\left[\phi_s - (\phi_s - \phi)^{1/2}\right] - 2J_e\sqrt{\frac{m}{2e}}\phi^{1/2}.$$

For space-charge-limited current flow, the electric field at the right-hand boundary (the edge of the plasma) is zero and the potential is $\phi = \phi_s$. Putting that into eq.

$$\frac{\varepsilon_o}{2}E^2 = 2J_i\sqrt{\frac{M}{2e}}\left[\phi_s - (\phi_s - \phi)^{1/2}\right] - 2J_e\sqrt{\frac{m}{2e}}\phi^{1/2}.$$

and solving for the current density gives:

$$J_e = \sqrt{\frac{M}{m}}J_i.$$

If the area of the two plasmas in contact with each other is the same, the electron current crossing the double layer is the square root of the mass ratio times the ion current crossing the layer. This situation is called the Langmuir condition (1929) and describes the space-charge-limited flow of ions and electrons between two plasmas or between a plasma and an electron emitter.

For finite initial velocities, Eq. $J_e = \sqrt{\frac{M}{m}}J_i.$ was corrected by Andrews and Allen to give:

$$J_e = \kappa\sqrt{\frac{M}{m}}J_i,$$

where κ is a constant that varies from 0.8 to 0.2 for T_e/T_i changing from 2 to about 20. For typical thruster plasmas where $T_e/T_i \approx 10$, k is about 0.5.

While the presence of free-standing double layers in the plasma volume in thrusters is often debated, the sheath at a thermionic cathode surface certainly satisfies the criteria of counter-streaming ion and electron currents and can be viewed as a double layer. In this case, Eq. $J_e = \kappa \sqrt{\dfrac{M}{m}} J_i$, describes the space-chargelimited current density that a plasma can accept from an electron-emitting cathode surface. This is useful in that the maximum current density that can be drawn from a cathode can be evaluated if the plasma parameters at the sheath edge in contact with the cathode are known (such that J_i can be evaluated from the Bohm current), without requiring that the actual sheath thickness be known.

Finally, there are several conditions for the formation of the classic double layer described here. In order to achieve a potential difference between the plasmas that is large compared to the local electron temperature, charge separation must occur in the layer. This, of course, violates quasi-neutrality locally. The current flow across the layer is space-charge limited, which means that the electric field is essentially zero at both boundaries. Finally, the flow through the layer discussed here is collisionless. Collisions cause resistive voltage drops where current is flowing, which can easily be confused with the potential difference across a double layer.

MAGNETOHYDRODYNAMICS

Magnetohydrodynamics (MHD) (magnetofluiddynamics or hydromagnetics) is the academic discipline which studies the dynamicsof electrically conducting fluids. Examples of such fluids include plasmas, liquid metals, and salt water. The word magnetohydrodynamics (MHD) is derived from magneto- meaning magnetic field, and hydro- meaning liquid, and –dynamics meaning movement. The field of MHD was initiated by Hannes Alfvén, for which he received the Nobel Prize in Physics in 1970.

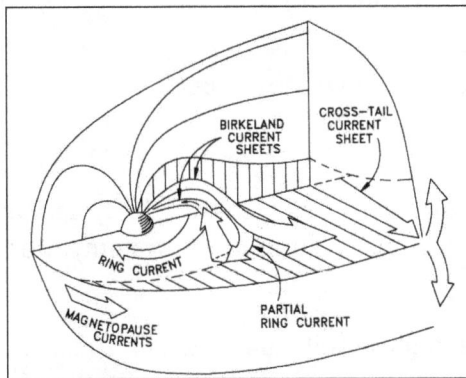

Schematic view of the different current systems
which shape the Earth's magnetosphere.

The idea of MHD is that magnetic fields can induce currents in a moving conductive fluid, which create forces on the fluid, and also change the magnetic field itself. The set of equations which describe MHD are a combination of the Navier-Stokes equations of fluid dynamics and Maxwell's equations of electromagnetism. These differential equations have to be solved simultaneously, either analytically or numerically. Because MHD is a fluid theory, it cannot treat kinetic phenomena, i.e., those in which the existence of discrete particles, or of a non-thermal distribution of their velocities, is important.

Ideal and Resistive MHD

The simplest form of MHD, Ideal MHD, assumes that the fluid has so little resistivity that it can be treated as a perfect conductor. In ideal MHD, Lenz's law dictates that the fluid is in a sense *tied* to the magnetic field lines. To be more precise, in ideal MHD, a small rope-like volume of fluid surrounding a field line will continue to lie along a magnetic field line, even as it is twisted and distorted by fluid flows in the system.

The connection between magnetic field lines and fluid in ideal MHD fixes the topology of the magnetic field in the fluid — for example, if a set of magnetic field lines are tied into a knot, then they will remain so as long as the fluid/plasma has negligible resistivity. This difficulty in reconnecting magnetic field lines makes it possible to store energy by moving the fluid or the source of the magnetic field. The energy can then become available if the conditions for ideal MHD break down, allowing magnetic reconnection that releases the stored energy from the magnetic field.

Ideal MHD Equations

The ideal MHD equations consist of the continuity equation (mass), the momentum equation, Ampere's Law in the limit of no electric field and no electron diffusivity, and a temperature evolution equation. As with any fluid description to a kinetic system, a closure approximation must be applied to highest moment of the particle distribution equation. This is often accomplished with approximations to the heat flux through a condition of adiabaticity or isothermality.

Applicability of Ideal MHD to Plasmas

Ideal MHD is only strictly applicable when:

- The plasma is strongly collisional, so that the time scale of collisions is shorter than the other characteristic times in the system, and the particle distributions are therefore close to Maxwellian.

- The resistivity due to these collisions is small. In particular, the typical magnetic diffusion times over any scale length present in the system must be longer than any time scale of interest.

- We are interested in length scales much longer than the ion skin depth and Larmor radius perpendicular to the field, long enough along the field to ignore

Landau damping, and time scales much longer than the ion gyration time (system is smooth and slowly evolving).

Importance of Resistivity

In an imperfectly conducting fluid, the magnetic field can generally move through the fluid, following a diffusion law with the resistivity of the plasma serving as a diffusion constant. This means that solutions to the ideal MHD equations are only applicable for a limited time for a region of a given size before diffusion becomes too important to ignore. One can estimate the diffusion time across a Solar active region (from collisional resistivity) to be hundreds to thousands of years, much longer than the actual lifetime of a sunspot — so it would seem reasonable to ignore the resistivity. By contrast, a meter-sized volume of seawater has a magnetic diffusion time measured in milliseconds.

Even in physical systems which are large and conductive enough that simple estimates suggest that we can ignore the resistivity, resistivity may still be important: many instabilities exist that can increase the effective resistivity of the plasma by factors of more than a billion. The enhanced resistivity is usually the result of the formation of small scale structure like current sheets or fine scale magnetic turbulence, introducing small spatial scales into the system over which ideal MHD is broken and magnetic diffusion can occur quickly. When this happens, Magnetic Reconnection may occur in the plasma to release stored magnetic energy as waves, bulk mechanical acceleration of material, particle acceleration, and heat. Magnetic reconnection in highly conductive systems is important because it concentrates energy in time and space, so that gentle forces applied to a plasma for long periods of time can cause violent explosions and bursts of radiation.

When the fluid cannot be considered as completely conductive, but the other conditions for ideal MHD are satisfied, it is possible to use an extended model called resistive MHD. This includes an extra term in Ampere's Law which models the collisional resistivity. Generally MHD computer simulations are at least somewhat resistive because their computational grid introduces a numerical resistivity.

Importance of Kinetic Effects

Another limitation of MHD (and fluid theories in general) is that they depend on the assumption that the plasma is strongly collisional, so that the time scale of collisions is shorter than the other characteristic times in the system, and the particle distributions are Maxwellian. This is usually not the case in fusion, space and astrophysical plasmas. When this is not the case, or we are interested in smaller spatial scales, it may be necessary to use a kinetic model which properly accounts for the non-Maxwellian shape of the distribution function. However, because MHD is very simple, and captures many of the important properties of plasma dynamics, it is often qualitatively accurate, and is almost invariably the first model tried.

Effects which are essentially kinetic and not captured by fluid models include double layers, a wide range of instabilities, chemical separation in space plasmas and electron runaway.

Structures in MHD Systems

In many MHD systems, most of the electric current is compressed into thin, nearly-two-dimensional ribbons termed current sheets. These can divide the fluid into magnetic domains, inside of which the currents are relatively weak. Current sheets in the solar corona are thought to be between a few meters and a few kilometers in thickness, which is quite thin compared to the magnetic domains (which are thousands to hundreds of thousands of kilometers across). Another example is in the earth's magnetosphere, where current sheets separate topologically distinct domains, isolating most of the earth's ionosphere from the solar wind.

Extensions to Magnetohydrodynamics

Resistive MHD

Resistive MHD describes magnetized fluids with non-zero electron diffusivity. This diffusivity leads to a breaking in the magnetic topology.

Extended MHD

Extended MHD describes a class of phenomena in plasmas that are higher order than resistive MHD, but which can adequately be treated with a single fluid description. These include the effects of Hall physics, electron pressure gradients, finite Larmor Radii in the particle gyromotion, and electron inertia.

Two-Fluid MHD

Two-Fluid MHD describes plasmas that include a non-negligible electric field. As a result, the electron and ion momenta must be treated separately. This description is more closely tied to Maxwell's equations as an evolution equation for the electric field exists.

Hall MHD

In 1960, M. J. Lighthill criticized the applicability of ideal or resistive MHD theory for plasmas. It concerned the neglect of the "Hall current term", a frequent simplification made in magnetic fusion theory. Hall-magnetohydrodynamics (HMHD) takes into account this electric field description of magnetohydrodynamics.

Significant Aspects of Plasma Physics

Some of the fundamental concepts that are studied under plasma physics are surface-wave-sustained discharge, plasma stability, Thomson scattering, plasma parameters, corona discharge, coronal seismology, diffusion damping, double layer, etc. This chapter closely examines these fundamental concepts of plasma physics to provide an extensive understanding of the subject.

DUSTY PLASMA

A dusty plasma (or complex plasma) is a plasma containing nanometer or micrometer-sized particles suspended in it. Dust particles may be charged and the plasma and particles behave as a plasma , following electromagnetic laws for particle up to about 10 nm (or 100 nm if large charges are present). Dust particles may acrete into larger particles resulting in "grain plasmas". Dusty plasmas can also be the dominant current carrier. They are of special interest, since they can form liquid and crystalline states, plasma crystals, and the dynamics of the charged dust grains are directly observable.

Examples of dusty plasmas include comets, planetary rings, exposed dusty surfaces, and the zodiacal dust cloud and dust in interplanetary space, interstellar and circumstellar clouds.

Dusty plasmas are interesting because presence of particles significantly alters the charged particle equilibrium leading to different phenomena. Electrostatic coupling between the grains can vary over a wide range so that the states of the dusty plasma can change from weakly coupled (gaseous) to crystalline. Such plasmas are of interest as a non-Hamiltonian system of interacting particles and as a means to study generic fundamental physics of self-organization, pattern formation, phase transitions, and scaling.

The temperature of dust in a plasma may be quite different from its environoment. For example:

Dust plasma component	Temperature
Dust temperature	10 K
Molecular temperature	100 K
Ion temperature	1,000 K
Electron temperature	10,000 K

The electric potential of dust particles is typically 1–10 V (positive or negative). The potential is usually negative because the electrons are more mobile than the ions. The physics is essentially that of a Langmuir probe that draws no net current, including formation of a Debye sheath with a thickness of a few times the Debye length. If the electrons charging the dust grains are relativistic, then the dust may charge to several kilovolts. Field emission, which tends to reduce the negative potential, can be important due to the small size of the particles. The photoelectric effect and the impact of positive ions may actually result in a positive potential of the dust particles.

Dynamics

The motion of solid particles in a plasma follows the *momentum equation* for ions and electrons:

$$m\frac{dv}{dt} = mg + q(E+v\times B) - mv_c v + f$$

where m, q are the mass and charge of the particle, g is the gravitation acceleration, $mv_c v$ is due to viscosity, and f respresents all other forces including radiation pressure. q (E + v x B) is the Lorentz force, where E is the electric field, v is the velocity and B is the magnetic field.

Then depending in the size of the particle, there are four categories:

- Very small particles, where q (E + v × B) dominates over mg.

- Small grains, where q/m ≈ √G, and plasma still plays a major role in the dynamics.

- Large grains, where the electromagnetic term is negligible, and the particles are referred to as grains. Their motion is determined by gravity and viscosity, and the equation of motion becomes $mv_c v$ = mg.

- Large solid bodies. In centimeter and meter-sized bodies, viscosity may cause significant perturbations that can change an orbit. In kilometer-sized (or more) bodies, gravity and inertia dominate the motion.

ASTROPHYSICAL PLASMA

Astrophysical plasma is plasma outside of the solar system. It is studied as part of astrophysics and is commonly observed in space. The accepted view of scientists is that much of the baryonic matter in the universe exists in this state.

When matter becomes sufficiently hot, it becomes ionized and forms a plasma. This process breaks matter into its constituent particles which includes negatively-charged electrons and positively-charged ions. These electrically-charged particles are susceptible to influences by local electromagnetic fields. This includes strong fields generated by stars, and weak fields which exist in star forming regions, in interstellar space, and in intergalactic space. Similarly, electric fields are observed in some stellar astrophysical phenomena, but they are inconsequential in very low-density gaseous mediums.

Astrophysical plasma is often differentiated from space plasma, which typically refers to the plasma of the Sun, the solar wind, and the ionospheres and magnetospheres of the Earth and other planets.

Observing and Studying Astrophysical Plasma

Plasmas in stars can both generate and interact with magnetic fields, resulting in a variety of dynamic astrophysical phenomena. These phenomena are sometimes observed in spectra due to the Zeeman effect. Other forms of astrophysical plasmas can be influenced by preexisting weak magnetic fields, whose interactions may only be determined directly by polarimetry or other indirect methods. In particular, the intergalactic medium, the interstellar medium, the interplanetary medium and solar winds consist of diffuse plasmas.

Astrophysical plasma may also be studied in a variety of ways as they emit electromagnetic radiation across a wide range of the electromagnetic spectrum. Because astrophysical plasmas are generally hot, electrons in the plasmas are continually emitting X-rays through the process called bremsstrahlung. This radiation may be detected with X-ray telescopes located in the upper atmosphere or in space. Astrophysical plasmas also emit radio waves and gamma rays.

Possible Related Phenomena

Scientists are interested in active galactic nuclei because such astrophysical plasmas could be directly related to the plasmas studied in laboratories. Many of these phenomena seemingly exhibit an array of complex magnetohydrodynamic behaviors, such as turbulence and instabilities. Although these phenomena may occur on astronomical scales as large as the galactic core, many astrophysicists suggest that they do not significantly involve plasma effects but are caused by matter consumed by super massive black holes.

In Big Bang cosmology, the entire universe was in a plasma state prior to recombination. Afterwards, much of the universe reionized after the first quasars formed.

Studying astrophysical plasmas is part of mainstream academic astrophysics. Though plasma processes are part of the standard cosmological model, current theories indicate that they might have only a minor role to play in forming the very largest structures, such as voids, galaxy clusters and superclusters.

NON-NEUTRAL PLASMA

A non-neutral plasma is a plasma whose net charge creates an electric field large enough to play an important or even dominant role in the plasma dynamics. The simplest non-neutral plasmas are plasmas consisting of a single charge species. Examples of single species non-neutral plasmas that have been created in laboratory experiments are plasmas consisting entirely of electrons, pure ion plasmas, positron plasmas, and antiproton plasmas.

Non-neutral plasmas are used for research into basic plasma phenomena such as cross-magnetic field transport, nonlinear vortex interactions, and plasma waves and instabilities. They have also been used to create cold neutral antimatter, by carefully mixing and recombining cryogenic pure positron and pure antiproton plasmas.

Positron plasmas are also used in atomic physics experiments that study the interaction of antimatter with neutral atoms and molecules. Cryogenic pure ion plasmas have been used in studies of strongly coupled plasmas and quantum entanglement. More prosaically, pure electron plasmas are used to produce the microwaves in microwave ovens, via the magnetron instability.

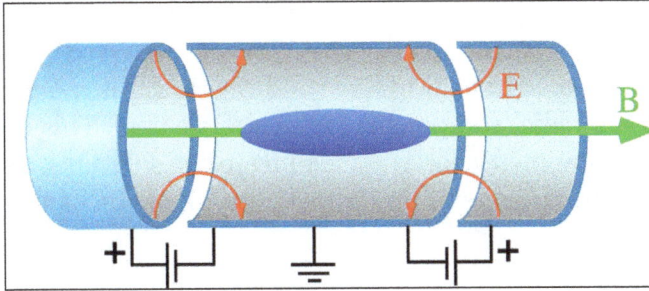

Neutral plasmas in contact with a solid surface (that is, most laboratory plasmas) are typically non-neutral in their edge regions. Due to unequal loss rates to the surface for electrons and ions, an electric field (the "ambipolar field") builds up, acting to hold back the more mobile species until the loss rates are the same. The electrostatic potential (as measured in electron-volts) required to produce this electric field depends on many variables but is often on the order of the electron temperature.

Non-neutral plasmas for which all species have the same sign of charge have exceptional confinement properties compared to neutral plasmas. They can be confined in a thermal equilibrium state using only static electric and magnetic fields, in a Penning trap configuration. Confinement times of up to several hours have been achieved. Using the "rotating wall" method, the plasma confinement time can be increased arbitrarily.

Such non-neutral plasmas can also access novel states of matter. For instance, they can be cooled to cryogenic temperatures without recombination (since there is no oppositely charged species with which to recombine). If the temperature is sufficiently low (typically on the order of 10 mK), the plasma can become a non-neutral liquid or a crystal. The body-centered-cubic structure of these plasma crystals has been observed by Bragg scattering in experiments on laser-cooled pure beryllium plasmas.

Equilibrium of a Single Species Non-neutral Plasma

Non-neutral plasmas with a single sign of charge can be confined for long periods of time using only static electric and magnetic fields. One such configuration is called a Penning trap, after the inventor F. M. Penning. The cylindrical version of the trap is also sometimes referred to as a Penning-Malmberg trap, after Prof. John Malmberg. The trap consists of several cylindrically symmetric electrodes and a uniform magnetic field applied along the axis of the trap. Plasmas are confined in the axial direction by biasing the end electrodes so as to create an axial potential well that will trap charges of a given sign (the sign is assumed to be positive in the figure). In the radial direction, confinement is provided by the v × B Lorentz force due to rotation of the plasma about the trap axis. Plasma rotation causes an inward directed Lorentz force that just balances the outward directed forces caused by the unneutralized plasma as well as the centrifugal force. Mathematically, radial force balance implies a balance between electric, magnetic and centrifugal forces:

$$0 = qE_r + qv_\theta B + mv_\theta^2 / r,$$

where particles are assumed to have mass m and charge q, r is radial distance from the trap axis and E_r is the radial component of the electric field. This quadratic equation can be solved for the rotational velocity v_θ, leading to two solutions, a slow-rotation and a fast-rotation solution. The rate of rotation $\omega = -v_\theta / r$ for these two solutions can be written as,

$$\omega = \frac{\Omega_c}{2} \pm \sqrt{\Omega_c^2 / 4 - qE_r / mr},$$

where $\Omega_c = qB / m$ is the cyclotron frequency. Depending on the radial electric field, the solutions for the rotation rate fall in the range $0 \le \omega / \Omega_c \le 1$. The slow and fast rotation modes meet when the electric field is such that $qE_r / mr = \Omega_c^2 / 4$. This is called the Brillouin limit; it is an equation for the maximum possible radial electric field that allows plasma confinement.

This radial electric field can be related to the plasma density n through the Poisson equation,

$$\frac{1}{r}\frac{\partial}{\partial r}(rE_r) = qn / \epsilon_0,$$

and this equation can be used to obtain a relation between the density and the plasma rotation rate. If we assume that the rotation rate is uniform in radius (i.e. the plasma rotates as a rigid body), then eq. $0 = qE_r + qv_\theta B + mv_\theta^2 / r$, implies that the radial electric field is proportional to radius r. Solving for E_r from this equation in terms of E_r and substituting the result into Poisson's equation yields:

$$n = \frac{2\epsilon_0 m\omega(\Omega_c - \omega)}{q^2}.$$

This equation implies that the maximum possible density occurs at the Brillouin limit, and has the value:

$$n_B = \frac{\epsilon_0 m\Omega_c^2}{2q^2} = \frac{B^2 / (2\mu_0)}{mc^2},$$

where $c = 1 / \sqrt{\mu_0 \epsilon_0}$ is the speed of light. Thus, the rest energy density of the plasma, n·m·c², is less than or equal to the magnetic energy density $B^2 / (2\mu_0)$ of the magnetic field. This is a fairly stringent requirement on the density. For a magnetic field of 10 tesla, the Brillouin density for electrons is only $n_B = 4.8 \times 10^{14}$ cm^{-3}.

The density predicted by eq. $n = \dfrac{2\epsilon_0 m\omega(\Omega_c - \omega)}{q^2}$, scaled by the Brillouin density, is shown as a function of rotation rate. Two rotation rates yield the same density, corresponding to the slow and fast rotation solutions.

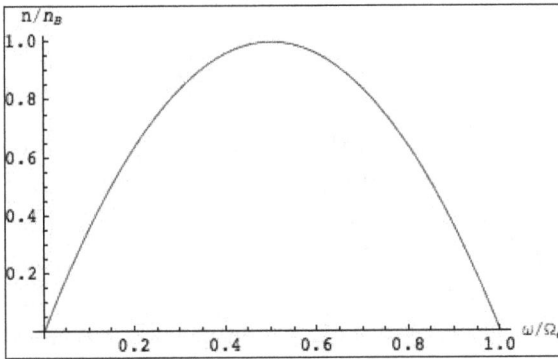

Density versus rotation rate for a single
species plasma confined in a Penning trap.

Plasma Loss Processes: The Rotating Wall Method

In experiments on single species plasmas, plasma rotation rates in the tens of kHz range are not uncommon, even in the slow rotation mode. This rapid rotation is necessary to provide the confining radial Lorentz force for the plasma. However, if there is neutral gas in the trap, collisions between the plasma and the gas cause the plasma rotation to slow, leading to radial expansion of the plasma until it comes in contact with the surrounding electrodes and is lost. This loss process can be alleviated by operating the trap in an ultra high vacuum. However, even under such conditions the plasma rotation can still be slowed through the interaction of the plasma with "errors" in the external confinement fields. If these fields are not perfectly cylindrically symmetric, the asymmetries can torque on the plasma, reducing the rotation rate. Such field errors are unavoidable in any actual experiment, and limit the plasma confinement time.

It is possible to overcome this plasma loss mechanism by applying a rotating field error to the plasma. If the error rotates faster than the plasma, it acts to spin up the plasma (similar to how the spinning blade of a blender causes the food to spin), counteracting the effect of field errors that are stationary in the frame of the laboratory. This rotating field error is referred to as a "rotating wall", after the theory idea that one could reverse the effect of a trap asymmetry by simply rotating the entire trap at the plasma rotation frequency. Since this is impractical, one instead rotates the trap electric field rather than the entire trap, by applying suitably phased voltages to a set of electrodes surrounding the plasma.

Cryogenic Non-neutral Plasmas: Correlated States

When a non-neutral plasma is cooled to cryogenic temperatures, it does not recombine to a neutral gas as would a neutral plasma, because there are no oppositely charged particles with which to recombine. As a result, the system can access novel strongly coupled non-neutral states of matter, including plasma crystals consisting solely of a single charge species. These strongly coupled non-neutral plasmas are parametrized by the coupling parameter Γ, defined as,

$$\Gamma = \frac{q^2}{4\pi\epsilon_0 k_B T a},$$

where T is the temperature and α is the Wigner-Seitz radius (or mean inter-particle spacing), given in terms of the density n by the expression $4\pi a^3 n / 3 = 1$. The coupling parameter can be thought of as the ratio of the mean interaction energy between nearest-neighbor pairs, $q^2 / (4\pi\epsilon_0 a)$, and the mean kinetic energy of order $k_b T$. When this ratio is small, interactions are weak and the plasma is nearly an ideal gas of charges moving in the mean-field produced by the other charges. However, when r> 1 interactions between particles are important and the plasma behaves more like a liquid, or even a crystal if is sufficiently large. In fact, computer simulations and theory have predicted that for an infinite homogeneous plasma the system exhibits a gradual onset of short-range order consistent with a liquid-like state for $\Gamma \approx 2$, and there is predicted to be a first-order phase transition to a body-centered-cubic crystal for $\Gamma \simeq 175$.

Experiments have observed this crystalline state in a pure beryllium ion plasma that was laser-cooled to the millikelvin temperature range. The mean inter-particle spacing in this pure ion crystal was on the order of 10-20 μm, much larger than in neutral crystalline matter. This spacing corresponds to a density on the order of 10^8-10^9 cm^{-3}, somewhat less than the Brillouin limit for beryllium in the 4.5 tesla magnetic field of the experiment. Cryogenic temperatures were then required in order to obtain a r value in the strongly coupled regime. The experiments measured the crystal structure by the Bragg-scattering technique, wherein a collimated laser beam was scattered off of the crystal, displaying Bragg peaks at the expected scattering angles for a bcc lattice.

When small numbers of ions are laser-cooled, they form crystalline "Coulomb clusters". The symmetry of the cluster depends on the form of the external confinement fields.

INDUCTION PLASMA

The 1960s were the incipient period of thermal plasma technology, spurred by the needs of aerospace programs. Among the various methods of thermal plasma generation, induction plasma (or inductively coupled plasma) takes up an important role.

Early attempts to maintain inductively coupled plasma on a stream of gas date back to Babat in 1947 and Reed in 1961. Effort was concentrated on the fundamental studies of energy coupling mechanism and the characteristics of the flow, temperature and concentration fields in plasma discharge. In 1980s, there was increasing interest in high-performance materials and other scientific issues, and in induction plasma for industrial-scale applications such as waste treatment. Numerous research and development were devoted to bridge the gap between the laboratory gadget and the industry integration. After decades' effort, induction plasma technology has gained a firm foothold in modern advanced industry.

Generation of Induction Plasma

Induction heating is a mature technology with centuries of history. A conductive metallic piece, inside a coil of high frequency, will be "induced", and heated to the red-hot state. There is no difference in cardinal principle for either induction heating or "inductively coupled plasma", only that the medium to induce, in the latter case, is replaced by the flowing gas, and the temperature obtained is extremely high, as it arrives the "fourth state of the matter"—plasma.

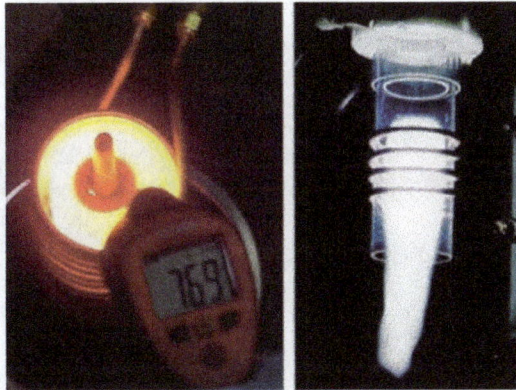

(left) Induction heating; (right) Inductively coupled plasma.

An inductively coupled plasma (ICP) torch is essentially a copper coil of several turns, through which cooling water is running in order to dissipate the heat produced in operation. The coil wraps a confinement tube, inside which the induction plasma is generated. One end of the confinement tube is open; the plasma is actually maintained on a continuum gas flow. During induction plasma operation, the generator supplies an alternating current (ac) of radio frequency (r.f.) to the torch coil; this ac induces an alternating magnetic field inside the coil, after Ampère's law (for a solenoid coil):

$$\phi_B = (\mu_0 I_c N)(\pi r_0^2)$$

where, ϕ_B is the flux of magnetic field, μ_0 is permeability constant $4\pi 10^{-7}$ Wb/A.m I_c is the coil current, N is the number of coil turns per unit length, and r_o is the mean radius of the coil turns.

According to Faraday's Law, a variation in magnetic field flux will induce a voltage, or electromagnetic force:

$$E = -N(\Delta\phi_B / \Delta t)$$

where, N is the number of coil turns, and the item in parenthesis is the rate at which the flux is changing. The plasma is conductive (assuming a plasma already exists in the torch). This electromagnetic force, E, will in turn drive a current of density j in closed loops. The situation is much similar to heating a metal rod in the induction coil: energy transferred to the plasma is dissipated via Joule heating, j^2R, from Ohm's law, where R is the resistance of plasma.

Since the plasma has a relatively high electrical conductivity, it is difficult for the alternating magnetic field to penetrate it, especially at very high frequencies. This phenomenon is usually described as the "skin effect". The intuitive scenario is that the induced currents surrounding each magnetic line counteract each other, so that a net induced current is concentrated only near the periphery of plasma. It means the hottest part of plasma is off-axis. Therefore, the induction plasma is something like an "annular shell". Observing on the axis of plasma, it looks like a bright "bagel".

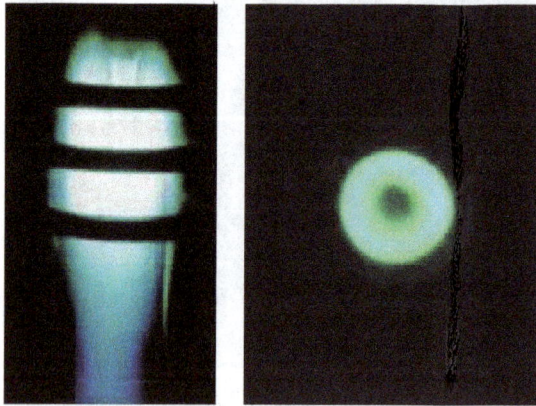

Induction plasma, observed from side and from the end.

In practice, the ignition of plasma under low pressure conditions (<300 torr) is almost spontaneous, once the r.f. power imposed on the coil achieves a certain threshold value (depending on the torch configuration, gas flow rate etc.). The state of plasma gas (usually argon) will swiftly transit from glow-discharge to arc-break and create a stable induction plasma. For the case of atmospheric ambient pressure conditions, ignition is often accomplished with the aid of a Tesla coil, which produces high-frequency, high-voltage electric sparks that induce local arc-break inside the torch and stimulate a cascade of ionization of plasma gas, ultimately resulting in a stable plasma.

Induction Plasma Torch

Induction plasma torch is the core of the induction plasma technology. Despite the existence of hundreds of different designs, an induction plasma torch consists of essentially three components:

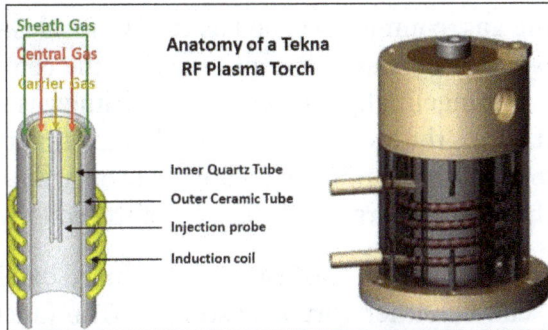

Induction plasma torch for industrial applications.

- Coil: The induction coil consists of several spiral turns, depending on the r.f. power source characteristics. Coil parameters including the coil diameter, number of coil turns, and radius of each turn, are specified in such a way to create an electrical "tank circuit" with proper electrical impedance. Coils are typically hollow along their cylindrical axis, filled with internal liquid cooling (e.g., de-ionized water) to mitigate high operating temperatures of the coils that result from the high electrical currents required during operation.

- Confinement tube: This tube serves to confine the plasma. Quartz tube is the common implementation. The tube is often cooled either by compressed air (<10 kW) or cooling water. While the transparency of quartz tube is demanded in many laboratory applications (such as spectrum diagnostic), its relatively poor mechanical and thermal properties pose a risk to other parts (e.g., o-ring seals) that may be damaged under the intense radiation of high-temperature plasma. These constraints limit the use of quartz tubes to low power torches only (<30 kW). For industrial, high power plasma applications (30~250 kW), tubes made of ceramic materials are typically used. The ideal candidate material will possess good thermal conductivity and excellent thermal shock resistance. For the time being, silicon nitride (Si_3N_4) is the first choice. Torches of even greater power employ a metal wall cage for the plasma confinement tube, with engineering tradeoffs of lower power coupling efficiencies and increased risk of chemical interactions with the plasma gases.

- Gas distributor: Often called a torch head, this part is responsible for the introduction of different gas streams into the discharge zone. Generally, there are three gas lines passing to the torch head. According to their distance to the center of circle, these three gas streams are also arbitrarily named as Q_1, Q_2, and Q_3.

Q_1 is the carrier gas that is usually introduced into the plasma torch through an injector at the center of the torch head. As the name indicates it, the function of Q_1 is to convey the precursor (powders or liquid) into plasma. Argon is the usual carrier gas, however, many other reactive gases (i.e., oxygen, NH_3, CH_4, etc.) are often involved in the carrier gas, depending on the processing requirement.

Q_2 is the plasma forming gas, commonly called as the "Central Gas". In today's induction plasma torch design, it is almost unexceptional that the central gas is introduced into the torch chamber by tangentially swirling. The swirling gas stream is maintained by an internal tube that hoops the swirl till to the level of the first turn of induction coil. All these engineering concepts are aiming to create the proper flow pattern necessary to insure the stability of the gas discharge in the center of the coil region.

Q_3 is commonly referred to as "Sheath Gas" that is introduced outside the internal tube. The flow pattern of Q_3 can be either vortex or straight. The function of sheath gas is twofold. It helps to stabilize the plasma discharge; most importantly, it protects the confinement tube, as a cooling medium.

- Plasma gases and plasma performance: The minimum power to sustain an induction plasma depends on pressure, frequency and gas composition. The lower sustaining power setting is achieved with high r.f. frequency, low pressure, and monatomic gas, such as argon. Once diatomic gas is introduced into the plasma, the sustaining power would be drastically increased, because extra dissociation energy is required to break gaseous molecular bonds first, so then further excitation to plasma state is possible. The major reasons to use diatomic gases in plasma processing are (1) to get a plasma of high energy content and good thermal conductivity, and (2) to conform the processing chemistry.

Gas	Specific gravity	Thermal dissocia- tion energy (eV)	Ionization energy (eV)	Thermal conductivity (W/m.K)	Enthalpy (MJ/mol)
Ar	1.380	not applicable	15.76	0.644	0.24
He	0.138	not applicable	24.28	2.453	0.21
H_2	0.069	4.59	13.69	3.736	0.91
N_2	0.967	9.76	14.53	1.675	1.49
O_2	1.105	5.17	13.62	1.370	0.99
Air	1.000	n.a.	n.a.	1.709	1.39

In practice, the selection of plasma gases in an induction plasma processing is first determined by the processing chemistry, i.e., if the processing requiring a reductive or oxidative, or other environment. Then suitable second gas may be selected and added to argon, so as to get a better heat transfer between plasma and the materials to treat. Ar–He, Ar–H_2, Ar–N_2, Ar–O_2, Air, etc. mixture are very commonly used induction plasmas. Since the energy dissipation in the discharge takes places essentially in the outer annular shell of plasma, the second gas is usually introduced along with the sheath gas line, rather than the central gas line.

Industrial Application of Induction Plasma Technology

Following the evolution of the induction plasma technology in laboratory, the major advantages of the induction plasma have been distinguished:

- Without the erosion and contamination concern of electrode, due to the different plasma generation mechanism compared with other plasma method, for example, direct current non-transfer arc (dc) plasma.

- The possibility of the axial feeding of precursors, being solid powders, or suspensions, liquids. This feature overcomes the difficulty of exposing materials to the high temperature of plasma, from the high viscosity of high temperature of plasma.

- Because of non electrode problem, a wide versatile chemistry selection is possible, i.e., the torch could work in either reductive, or, oxidative, even corrosive conditions. With this capability, induction plasma torch often works as not only a high temperature, high enthalpy heat source, but also chemical reaction vessels.

- Relatively long residence time of precursor in the plasma plume (several milliseconds up to hundreds milliseconds), compared with dc plasma.

- Relatively large plasma volume.

These features of induction plasma technology, has found niche applications in industrial scale operation in the last decade. The successful industrial application of induction plasma process depends largely on many fundamental engineering supports. For example, the industrial plasma torch design, which allows high power level (50 to 600 kW) and long duration (three shifts of 8 hours/day) of plasma processing. Another example is the powder feeders that convey large quantity of solid precursor (1 to 30 kg/h) with reliable and precise delivery performance.

Nowadays, we have been in a position to be able to numerate many examples of the industrial applications of induction plasma technology, such as, powder spheroidisation, nanosized powders synthesis, induction plasma spraying, waste treatments, etc., However, the most impressive success of induction plasma technology is doubtless in the fields of spheroidisation and nano-materials synthesis.

Powder Spheroidisation

The requirement of powders spheroidisation (as well as densification) comes from very different industrial fields, from powder metallurgy to the electronic packaging. Generally speaking, the pressing need for an industrial process to turn to spherical powders is to seek at least one of the following benefits which result from the spheroidisation process:

- Improve the powders flow-ability.

- Increase the powders packing density.

- Eliminate powder internal cavities and fractures.

- Change the surface morphology of the particles.

- Other unique motive, such as optical reflection, chemical purity etc.

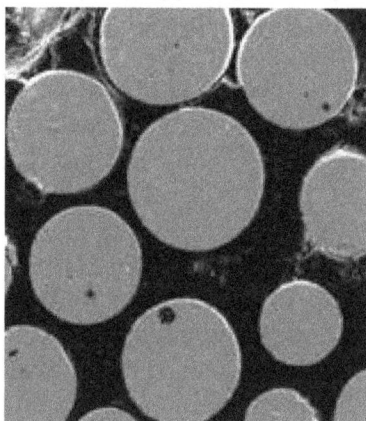

The dense microstructure of the spheroidised cast tungsten carbide powders.

Spheroidisation is a process of in-flight melting. The powder precursor of angular shape is introduced into induction plasma, and melted immediately in the high temperatures of plasma. The melted powder particles are assuming the spherical shape under the action of surface tension of liquid state. These droplets will be drastically cooled down when fly out of the plasma plume, because of the big temperature gradient exciting in the plasma. The condensed spheres are thus collected as the spheroidisation products.

A great variety of ceramics, metals and metal alloys have been successfully spheroidized/densified using induction plasma spheroidisation. Following are some typical materials spheroidized on commercial scale.

- Oxide ceramics: SiO_2, ZrO_2, YSZ, Al_2TiO_5, glass.

- Non-oxides: WC, WC–Co, CaF_2, TiN.

- Metals: Re, Ta, Mo, W.

- Alloys: Cr–Fe–C, Re–Mo, Re–W.

Nano-materials Synthesis

It is the increased demand for nanopowders that promotes the extensive research and development of various techniques for nanometric powders. The challenges for an industrial application technology are productivity, quality controllability, and affordability. Induction plasma technology implements in-flight evaporation of precursor, even those raw materials of the highest boiling point; operating under various atmospheres, permitting synthesis of a great variety of nanopowders, and thus become much more reliable and efficient technology for synthesis of nanopowders in both laboratory and industrial scales. Induction plasma used for nanopowder synthesis has many advantages over the alternative techniques, such as high purity, high flexibility, easy to scale up, easy to operate and process control.

In the nano-synthesis process, material is first heated up to evaporation in induction plasma, and the vapours are subsequently subjected to a very rapid quenching in the quench/reaction zone. The quench gas can be inert gases such as Ar and N_2 or reactive gases such as CH_4 and NH_3, depending on the type of nanopowders to be synthesized. The nanometric powders produced are usually collected by porous filters, which are installed away from the plasma reactor section. Because of the high reactivity of metal powders, special attention should be given to powder pacification prior to the removal of the collected powder from the filtration section of the process.

The induction plasma system has been successfully used in the synthesis nanopowders. The typical size range of the nano-particles produced is from 20 to 100 nm, depending on the quench conditions employed. The productivity varies from few hundreds g/h to 3~4 kg/h, according to the different materials' physical properties.

SURFACE-WAVE-SUSTAINED DISCHARGE

A surface-wave-sustained discharge is a plasma that is excited by propagation of electromagnetic surface waves. Surface wave plasma sources can be divided into two groups depending upon whether the plasma generates part of its own waveguide by ionisation or not. The former is called a self-guided plasma. The surface wave mode allows the generation of uniform high-frequency-excited plasmas in volumes whose lateral dimensions extend over several wavelengths of the electromagnetic wave, e.g. for microwaves of 2.45 GHz in vacuum the wavelength amounts to 12.2 cm.

Theory

For a long time, microwave plasma sources without a magnetic field were not considered suitable for the generation of high density plasmas. Electromagnetic waves cannot propagate in over-dense plasmas. The wave is reflected at the plasma surface due to the skin effect and becomes an evanescent wave. Its penetration depth corresponds to the skin depth , which can be approximated by,

$$\delta \approx c / \sqrt{\omega_{p_e}^2 - \omega^2}.$$

The non-vanishing penetration depth of an evanescent wave opens an alternative way of heating a plasma: Instead of *traversing* the plasma, the conductivity of the plasma enables the wave to propagate *along* the plasma surface. The wave energy is then transferred to the plasma by an evanescent wave which enters the plasma perpendicular to its surface and decays exponentially with the skin depth. 'Transfer mechanism allows to generate over-dense plasmas with electron densities beyond the critical density.

Design

Surface-wave-sustained plasmas (SWP) can be operated in a large variety of recipient geometries. The pressure range accessible for surface-wave-excited plasmas depends on the process gas and the diameter of the recipient. The larger the chamber diameter, the lower the minimal pressure necessary for the SWP mode. Analogously, the maximal pressure where a stable SWP can be operated decreases with increasing diameter.

The numerical modelling of SWPs is quite involved. The plasma is created by the electromagnetic wave, but it also reflects and guides this same wave. Therefore, a truly self-consistent description is necessary.

PLASMA PARAMETERS

Plasma parameters define various characteristics of a plasma, an electrically conductive collection of charged particles that responds *collectively* to electromagnetic forces. Plasma typically takes the form of neutral gas-like clouds or charged ion beams, but may also include dust and grains. The behaviour of such particle systems can be studied statistically.

Fundamental Plasma Parameters

All quantities are in Gaussian (cgs) units except energy and temperature expressed in eV and ion mass expressed in units of the proton mass $\mu = m_i / m_p$; Z is charge state; k is Boltzmann's constant; K is wavenumber; $\ln \Lambda$ is the Coulomb logarithm.

Frequencies

- Electron gyrofrequency, the angular frequency of the circular motion of an electron in the plane perpendicular to the magnetic field:

$$\omega_{ce} = \frac{eB}{m_e c} \approx 1.76 \times 10^7 \, B \text{ rad/s}$$

- Ion gyrofrequency, the angular frequency of the circular motion of an ion in the plane perpendicular to the magnetic field:

$$\omega_{ci} = \frac{ZeB}{m_i c} \approx 9.58 \times 10^3 \frac{ZB}{\mu} \text{ rad/s}$$

- Electron plasma frequency, the frequency with which electrons oscillate (plasma oscillation):

$$\omega_{pe} = \left(\frac{4\pi n_e e^2}{m_e}\right)^{\frac{1}{2}} \approx 5.64 \times 10^4 \, n_e^{\frac{1}{2}} \text{ rad/s}$$

- Ion plasma frequency:

$$\omega_{pi} = \left(\frac{4\pi n_i Z^2 e^2}{m_i}\right)^{\frac{1}{2}} \approx 1.32 \times 10^3 \, Z \left(\frac{n_i}{\mu}\right)^{\frac{1}{2}} \text{ rad/s}$$

- Electron trapping rate:

$$v_{Te} = \left(\frac{eKE}{m_e}\right)^{\frac{1}{2}} \approx 7.26 \times 10^8 \left(KE\right)^{\frac{1}{2}} / \text{s}$$

- Ion trapping rate:

$$v_{Ti} = \left(\frac{ZeKE}{m_i}\right)^{\frac{1}{2}} \approx 1.69 \times 10^7 \left(\frac{ZKE}{\mu}\right)^{\frac{1}{2}} / \text{s}$$

- Electron collision rate in completely ionized plasmas:

$$v_e \approx 2.91 \times 10^{-6} \frac{n_e \ln \Lambda}{T_e^{\frac{3}{2}}} / \text{s}$$

- Ion collision rate in completely ionized plasmas:

$$v_i \approx 4.80 \times 10^{-8} \frac{Z^4 n_i \ln \Lambda}{\left(T_i^3 \mu\right)^{\frac{1}{2}}} / \text{s}$$

Lengths

- Electron thermal de Broglie wavelength, approximate average de Broglie wavelength of electrons in a plasma:

$$\lambda_{th,e} = \sqrt{\frac{h^2}{2\pi m_e k T_e}} = 6.919 \times 10^{-8} \, T_e^{-1/2} \text{ cm}$$

- Classical distance of closest approach, the closest that two particles with the elementary charge come to each other if they approach head-on and each has a velocity typical of the temperature, ignoring quantum-mechanical effects:

$$\frac{e^2}{kT} \approx 1.44 \times 10^{-7} \frac{1}{T} \text{ cm}$$

- Electron gyroradius, the radius of the circular motion of an electron in the plane perpendicular to the magnetic field:

$$r_e = \frac{v_{Te}}{\omega_{ce}} \approx 2.38 \frac{T_e^{\frac{1}{2}}}{B} \text{ cm}$$

- Ion gyroradius, the radius of the circular motion of an ion in the plane perpendicular to the magnetic field:

$$r_i = \frac{v_{Ti}}{\omega_{ci}} \approx 1.02 \times 10^2 \frac{(\mu T_i)^{\frac{1}{2}}}{ZB} \text{ cm}$$

- Plasma skin depth (also called the electron inertial length), the depth in a plasma to which electromagnetic radiation can penetrate:

$$\frac{c}{\omega_{pe}} \approx 5.31 \times 10^5 \frac{1}{n_e^{\frac{1}{2}}} \text{ cm}$$

- Debye length, the scale over which electric fields are screened out by a redistribution of the electrons:

$$\lambda_D = \left(\frac{kT_e}{4\pi ne^2}\right)^{\frac{1}{2}} = \frac{v_{Te}}{\omega_{pe}} \approx 7.43 \times 10^2 \left(\frac{T_e}{n}\right)^{\frac{1}{2}} \text{ cm}$$

- Ion inertial length, the scale at which ions decouple from electrons and the magnetic field becomes frozen into the electron fluid rather than the bulk plasma:

$$d_i = \frac{c}{\omega_{pi}} \approx 2.28 \times 10^7 \frac{1}{Z} \left(\frac{\mu}{n_i}\right)^{\frac{1}{2}} \text{ cm}$$

- Mean free path, the average distance between two subsequent collisions of the electron (ion) with plasma components:

$$\lambda_{e,i} = \frac{\overline{v_{e,i}}}{\nu_{e,i}},$$

where $\overline{v}_{e,i}$ is an average velocity of the electron (ion) and $v_{e,i}$ is the electron or ion collision rate.

Velocities

- Electron thermal velocity, typical velocity of an electron in a Maxwell–Boltzmann distribution:

$$v_{Te} = \left(\frac{kT_e}{m_e} \right)^{\frac{1}{2}} \approx 4.19 \times 10^7 \, T_e^{\frac{1}{2}} \text{ cm/s}$$

- Ion thermal velocity, typical velocity of an ion in a Maxwell–Boltzmann distribution:

$$v_{Ti} = \left(\frac{kT_i}{m_i} \right)^{\frac{1}{2}} \approx 9.79 \times 10^5 \left(\frac{T_i}{\mu} \right)^{\frac{1}{2}} \text{ cm/s}$$

- Ion speed of sound, the speed of the longitudinal waves resulting from the mass of the ions and the pressure of the electrons:

$$c_s = \left(\frac{\gamma Z k T_e}{m_i} \right)^{\frac{1}{2}} \approx 9.79 \times 10^5 \left(\frac{\gamma Z T_e}{\mu} \right)^{\frac{1}{2}} \text{ cm/s}$$

where γ is the adiabatic index.

- Alfvén velocity, the speed of the waves resulting from the mass of the ions and the restoring force of the magnetic field:

$$v_A = \frac{B}{(4\pi n_i m_i)^{\frac{1}{2}}} \approx 2.18 \times 10^{11} \frac{B}{(\mu n_i)^{\frac{1}{2}}} \text{ cm/s in cgs units,}$$

$$v_A = \frac{B}{(\mu_0 n_i m_i)^{\frac{1}{2}}} \text{ in SI units.}$$

Dimensionless

- Number of particles in a Debye sphere:

$$\left(\frac{4\pi}{3} \right) n \lambda_D^3 \approx 1.72 \times 10^9 \left(\frac{T^3}{n} \right)^{\frac{1}{2}}$$

- Alfvén velocity/speed of light.

$$\frac{v_A}{c} \approx 7.28 \frac{B}{\left(\mu n_i\right)^{\frac{1}{2}}}$$

- Electron plasma/gyrofrequency ratio.

$$\frac{\omega_{pe}}{\omega_{ce}} \approx 3.21 \times 10^{-3} \frac{n_e^{\frac{1}{2}}}{B}$$

- Ion plasma/gyrofrequency ratio.

$$\frac{\omega_{pi}}{\omega_{ci}} \approx 0.137 \frac{\left(\mu n_i\right)^{\frac{1}{2}}}{B}$$

- Thermal/magnetic pressure ratio ("beta").

$$\beta = \frac{8\pi nkT}{B^2} \approx 4.03 \times 10^{-11} \frac{nT}{B^2}$$

- Magnetic/ion rest energy ratio.

$$\frac{B^2}{8\pi n_i m_i c^2} \approx 26.5 \frac{B^2}{\mu n_i}$$

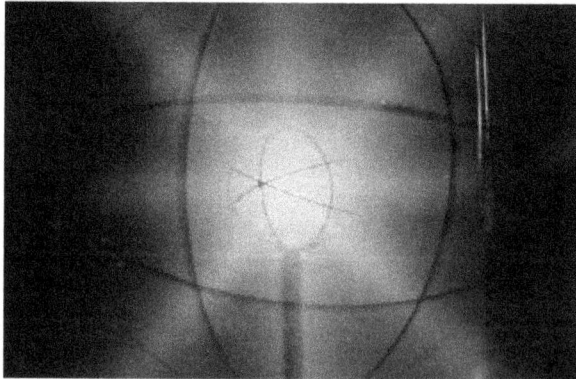

A 'sun in a test tube'. The Farnsworth-Hirsch Fusor during operation in
so called "star mode" characterized by "rays" of glowing plasma which appear
to emanate from the gaps in the inner grid.

Collisionality

In the study of tokamaks, collisionality is a dimensionless parameter which expresses the ratio of the electron-ion collision frequency to the banana orbit frequency.

The plasma collisionality v^* is defined as,

$$v^* = v_{ei}\sqrt{\frac{m_e}{k_B T_e}}\epsilon^{-3/2}qR,$$

where R denotes the electron-ion collision frequency, ϵ is the major radius of the plasma, is the inverse aspect-ratio, and q is the safety factor. The plasma parameters m_i and T_i denote, respectively, the mass and temperature of the ions, and k_B is the Boltzmann constant.

DOUBLE LAYER

A double layer is a structure in a plasma and consists of two parallel layers with opposite electrical charge. The sheets of charge cause a strong electric field and a correspondingly sharp change in voltage (electrical potential) across the double layer. Ions and electrons which enter the double layer are accelerated, decelerated, or reflected by the electric field. In general, double layers (which may be curved rather than flat) separate regions of plasma with quite different characteristics. Double layers are found in a wide variety of plasmas, from discharge tubes to space plasmas to the Birkeland currents supplying the Earth's aurora, and are especially common in current-carrying plasmas. Compared to the sizes of the plasmas which contain them, double layers are very thin (typically ten Debye lengths), with widths ranging from a few millimeters for laboratory plasmas to thousands of kilometres for astrophysical plasmas.

Saturnian aurora whose reddish colour is
characteristic of ionized hydrogen plasma.

Powered by the Saturnian equivalent of (filamentary) Birkeland currents, streams of charged particles from the interplanetary medium interact with the planet›s magnetic field and funnel down to the poles. Double layers are associated with (filamentary) currents, and their electric fields accelerate ions and electrons.

Other names for a double layer are: electrostatic double layer, electric double layer, plasma double layers, electrostatic shock (a type of double layer which is oriented at an oblique angle to the magnetic field in such a way that the perpendicular electric field is much larger than the parallel electric field), space charge layer, and "potential ramp". In laser physics, a double layer is sometimes called an ambipolar electric field. Double layers are conceptually related to the concept of a 'sheath'.

The adopted electrical symbol for a double layer, when represented in an electrical circuit is:

————DL————

If there is a net current present, then the DL is oriented so that the base of the L is in line with direction of current.

Double Layer Classification

Hall effect thruster. The electric fields utilised in plasma thrusters (in particular the Helicon Double Layer Thruster) may be in the form of double layers. Double layers may be classified in the following ways:

- Weak and strong double layers: The strength of a double layer is expressed as the ratio of the potential drop in comparison with the plasma's equivalent thermal potential, or in comparison with the rest mass energy of the electrons. A double layer is said to be strong if the potential drop across the layer is greater than the equivalent thermal potential of the plasma's components. This means that for strong double layers there are four different components to the plasma: 1. the electrons entering at the low potential side of the double layer which are accelerated; 2. the ions entering at the high potential side of the double layer which are accelerated; 3. the electrons entering at the high potential side of the double layer which are decelerated and successively refelected; and 4. the ions which enter the double layer at the low potential side of the double layer which are decelerated and reflected. Note that in the case of a weak double layer, the electrons and ions entering from the "wrong" side are decelerated by the electric field, however most will not be reflected, as the potential drop is not strong enough.

- Relativistic or nonrelativistic double layers: A double layer is said to be relativistic if the potential drop over the layer is so large that the total gain in energy of the particles is larger than the rest mass energy of the electron. The charge distribution in a relativistic double layer is such that the charge density is located in two very thin layers, and inside the double layer the density is constant at and very low compared to the rest of the plasma. In this respect, the double layer is similar to the charge distribution in a capacitor. As a special case of a relativistic double layer one can consider the vacuum gap at the magnetic polar cap of a pulsar.

- Current carrying and current-free double layers may both occur: Current carrying double layers may be generated by current-driven plasma instabilities which amplify variations of the plasma density. Current-free double layers form on the interface between two plasma regions with different characteristics, and their associated electric field maintains a balance between the penetration of electrons in either direction (so that the net current is low).

Double Layer Formation

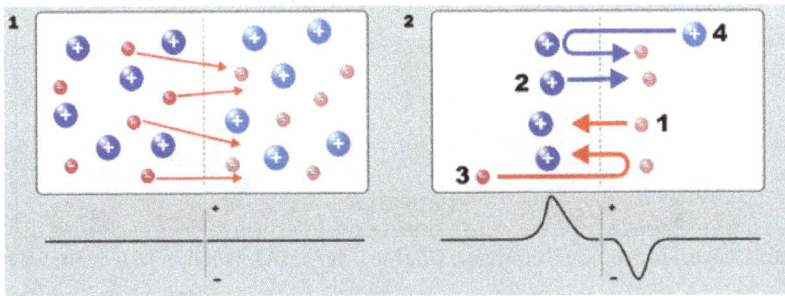

Doubler layer formation. Hotter electrons moving into a cooler plasma region cause a charge imbalance, resulting in a double layer that is able to accelerate electrons across it.

There are two different kinds of double layers, which are formed differently:

Current Carrying Double Layers

Current carrying double layers may arise in plasmas carrying a current. Various instabilities can be responsible for the formation of these layers. One example is the Buneman instability which occurs when the streaming velocity of the electrons (basically the current density divided by the electron density) exceeds the electron thermal velocity of the plasma. Double layers (and other phase space structures) are often formed in the non-linear phase of the instability. One way of viewing the Buneman instability is to describe what happens when the current (in the form of a zero temperature electron beam) has to pass through a region of decreased ion density. In order to prevent charge from accumulating, the current in the system must be the same everywhere (in this 1D model). The electron density also has to be close to the ion density (quasineutrality), so there is also a dip in electron density. The electrons must therefore be accelerated into the density cavity, to maintain the same current density with a lower density of charge carriers. This implies that the density cavity is at a high electrical potential. As a consequence, the ions are accelerated out of the cavity, amplifying the density perturbation. Then there is the situation of a double-double layer, of which one side will most likely be convected away by the plasma, leaving a regular double layer. This is the process in which double layers are produced along planetary magnetic field lines in so-called Birkeland currents.

Current-free Double Layers

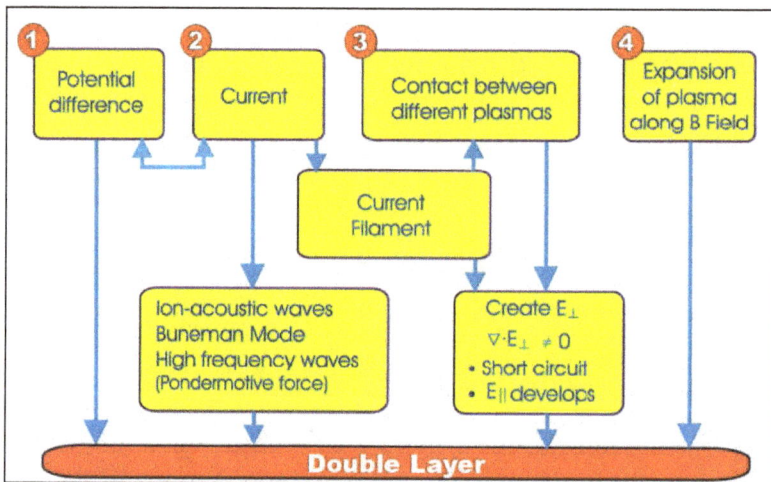

Double Layer Formation Summary. Double layers are formed in four main ways.

Current-free double layers occur at the boundary between plasma regions with different plasma properties. We explain how they form (neglecting the ions which are considered solely as a neutralizing background). Consider a plasma divided into two regions by a plane, which has a higher electron temperature on one side than on the other (the same analysis can also be done for different densities). This means that the electrons on one side of the interface have a greater thermal velocity. The electrons may stream freely in either direction, and the flux of electrons from the hot plasma to the cold plasma will be greater than the flux of the electrons from the cold plasma to the hot plasma, because the electrons from the hot side have a greater average speed. Because many more electrons enter the cold plasma than exit it, part of the cold region becomes negatively charged. The hot plasma, conversely, becomes positively charged. Therefore, an electric field builds up, which starts to accelerate electrons towards the hot region, reducing the net flux. In the end, the electric field builds up until the fluxes of electrons in either direction are equal, and further charge build up in the two plasmas is prevented. The potential drop is in fact exactly equal to the difference in thermal potential between the two plasma regions in this case, so such a double layer is a marginally strong double layer.

Double Layer Formation Mechanisms

The Moon. The prediction of a lunar
double layer was confirmed in 2003.

In the shadows, the Moon charges negatively in the interplanetary medium. and when the Moon passes through the Earth's Magnetotail.

The details of the formation mechanism depend on the environment of the plasma (eg. double layers in the laboratory, ionosphere, space plasmas, fusions plasma, etc). Proposed mechanisms for their formation have included:

- 1971: Between plasmas of different temperatures.

- 1976: In laboratory plasmas.

- 1982: Disruption of a neutral current sheet.

- 1983: Injection of non-neutral electron current into a cold plasma.

- 1985: Increasing the current density in a plasma.

- 1986: In the accretion column of a neutron star.

- 1986: By pinches in cosmic plasma regions.

- 1987: In a plasma constrained by a magnetic mirror.

- 1988: By an electrical discharge.

- 1988: Current-driven instabilities (strong double layers).

- 1988: Spacecraft-ejected electron beams.

- 1989: From shock waves in a plasma.

- 2000: Laser radiation.

- 2002: When magnetic field-aligned currents encounter density cavities.

- 2003: By the incidence of plasma on the dark side of the Moon's surface.

Features and Characteristics of Double Layers

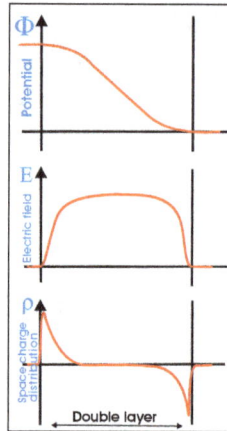

Double layer characteristics showing the potential (Φ),
electric field (E) and space charge distribution (ρ) across the layer.

- Thickness: The production of a double layer requires regions with a significant excess of positive or negative charge, that is, where quasi-neutrality is violated. In general, quasi-neutrality can only be violated on scales of the order of the Debye length. The thickness of a double layer is of the order of ten Debye lengths, which is a few centimeters in the ionosphere, a few tens of meters in the interplanetary medium, and tens of kilometers in the intergalactic medium.

- Particle acceleration: The potential drop across the double layer will accelerate electrons and positive ions in opposite directions. The magnitude of the potential drop determines the acceleration of the charged particles. In strong double layers, this will result in beams or jets of charged particles.

- Particle populations: As described in the formation of double layers, there are four populations of charge particles inside a double layer (1) Free electrons that are accelerated across the double layer (2) Free positive ions that are accelerated in the opposite direction across the double layer (3) Reflected electrons that approach the double layer, but are reflected back and counter stream away (4) Reflected positive ions that approach the double layer, but are reflected back and counter stream away. Note that in the case of weak double layers not all electrons and ions entering "from the wrong side" will be reflected, and therefore there will also be a population of decelerated electrons and ions.

- Particle flux: For non-relativistic current carrying double layers the electrons comprise the main part of the particle flux. The Langmuir condition states that the ratio of the electron and the ion current across the layer is given by the square root of the mass ratio of the ions to the electrons. For relativistic double layers the current ratio is 1, i.e. equal amounts of current are carried by the electrons and the ions.

- Energy supply: In a certain limit, the voltage drop across a current-carrying double layer is proportional to the total current, and it might be thought of as a resistive element (or *load*) which absorbs energy in an electric circuit. Anthony Peratt (1991) wrote: "Since the double layer acts as a load, there has to be an external source maintaining the potential difference and driving the current. In the laboratory this source is usually an electrical power supply, whereas in space it may be the magnetic energy stored in an extended current system, which responds to a change in current with an inductive voltage".

- Stability: Double layers in laboratory plasmas may be stable or unstable depending on the parameter regime. Various types of instabilities may occur, often arising due to the formation of beams of ions and electrons. Unstable double layers are *noisy* in the sense that they produce oscillations across a wide frequency band. A lack of plasma stability may also lead to a dramatic change in configuration often referred to as an explosion (and hence *exploding double layer*). In one example, the region enclosed in the double layer rapidly expands and evolves. An explosion of this type was first discovered in mercury rectifiers used in high-power direct-current transmission lines, where the voltage drop across the device was seen to increase by several orders of magnitude. Double layers may also drift, usually in the direction of the emitted electron beam, and in this respect are natural analogues to the smooth bore magnetron.

Table: Typical double layers.

Location	Typical Voltage drop	Source
Ionosphere	$10^2 - 10^4 V$	Satellite
Solar	$10^9 - 10^{11} V$	Estimated
Neutron star	$10^{15} V$	Estimated
Galactic filament	$10^{17} V$	Estimated

- Magnetised plasmas: Double layers can both form in unmagnetised and magnetised plasmas.

- Cellular nature: While double layers are relatively thin, they will spread over the entire cross surface of a laboratory container. Likewise where adjacent plasma regions have different properties, double layers will form and tend to cellularise the different regions.

- Energy transfer: Double layers fascilitate the transfer of electrical energy into kinetic energy, dW/dt=I.ΔV where I is the electric current dissipating energy into a double layer with a voltage drop of ΔV. Alfvén points out that the current may well consist exclusively of low-energy particles. Torvén *et al.* also report that plasma may spontaneously transfer magnetically stored energy into kinetic energy by electric double layers.

- Oblique double layer: An oblique double layer has its electric field not parallel to the background magnetic field (i.e. it is not field-aligned).

- Simulation: Double layers may be modelled using kinetic computer models like particle-in-cell (PIC) simulations. In some cases it is reasonable to treat the plasma as essentially one- or two-dimensional to reduce the computational cost of a simulation.

- Bohm criterion: A double layer cannot exist under all circumstances. In order to achieve that the electric field vanishes at the boundaries of the double layer, an existence criterion says that there is a maximum to the temperature of the ambient plasma. This is the so-called Bohm criterion.

- Bio-physical analogy: A model of plasma double layers has been used to investigate their applicability to understanding ion transport across biological cell membranes. Brazilian researchers have note that "Concepts like *charge neutrality*, *Debye length*, and *double layer* are very useful to explain the electrical properties of a cellular membrane". Plasma physicist Hannes Alfvén also noted that association of double layers with cellular structure, as had Irving Langmuir-before him, who coined the named "plasma" after its resemblance to blood cells.

The research of these objects is a relatively young phenomenon. Although it was already known in the 1920s that a plasma has a limited capacity for current maintenance Irving Langmuir. He characterized double layers in the laboratory and called these structures double-sheaths. It was not until the 1950s that a thorough study of double layers started in the laboratory. At the moment many groups are working on this topic theoretically, experimentally and numerically.

It was first proposed by Hannes Alfvén (the developer of magnetohydrodynamics) that the creation of the polar lights or Aurora Borealis is created by accelerated electrons in the magnetosphere of the Earth. He supposed that the electrons were accelerated electrostatically by an electric field localized in a small volume bounded by two charged regions. This so-called double layer would accelerate electrons Earthwards. Many experiments with rockets and satellites have been performed to investigate the magnetosphere and acceleration regions. The first indication for the existence of electric field aligned along the magnetic field (or double layers) in the magnetosphere was by a rocket experiment by McIlwain. Later, in 1977, Forrest Mozer reported that satellites had detected the signature of double layers (which he called electrostatic shocks) in the magnetosphere.

He most definite proof of these structures was obtained by the Viking satellite, which measures the differential potential structures in the magnetosphere with probes mounted on 40 m long booms. These probes can measure the local particle density and the potential difference between two points 80m apart. Asymmetric potential structures with respect to 0 V were measured, which means that the structure has a net potential and can be regarded as a double layer. The particle densities measured in such structures can be a los as 33% of the background density. The structures usually have an extent of 100 m (a few tens of Debye lengths). The filling factor of the lower magnetosphere with such solitary structures is about 10%. If one out of 5 such structures has a net potential drop of 1 V then the total potential drop over a region of 5000 km would be more than the 1 kV which is needed for the electrons to create the aurora. Magnetospheric double layers typically have a strength $e\phi_{DL}/k_B T_e \approx 0.1$ (where the electron temperature is assumed to lie in the range ($2eV \leq k_B T_e \leq 20eV$) and are therefore weak.

In the laboratory double layers can be created in different devices. The are investigated in double plasma machines, triple plasma machines and Q-machines. The stationary potential structures which can be measured in these machines agree very well with what one would expect theoretically. An example of a laboratory double layer can be seen in the figure below, taken from Torvén and Linberg, where we can see how well-defined and confined the potential drop of a double layer in a double plasma machine is.

On of the interesting things of the experiment by Torvén and Lindberg is that not only did the measure the potential structure in the double plasma machine but they also found high-frequency fluctuating electric fields at the high-potential side of the double layer. These fluctuations are probably due to a beam-plasma interaction outside the double layer which excites plasma turbulence. Their observations are consistent with experiments on electromagnetic radiation emitted by double layers in a double plasma machine by Volwerk, who, however, also observed radiation from the double layer itself. The power of these fluctuations has a maximum around the plasma frequency of the ambient plasma.

A recent development in double layer experiments is the investigation of so-called stairstep double layers. It has been observed that a potential drop in a plasma column can be split up into different parts. Transitions from a single double layer into two, three or more-step double layers are strongly sensitive to the boundary conditions of the plasma. These experiments can give us information about the formation of the magnetospheric double layers and their possible role in creating the aurora.

Some scientists have subsequently suggested a role of double layers in solar flares.

Mathematical Description of a Double Layer

In this topic we will take a closer look at the mathematics behind double layers. We first describe a semi-quantitative criterion for the formation of a density dip. We then describe a particularly simple kind of double layer. We then explain how to use the distribution function and the Vlasov-Poisson equation to model more complex double layers.

Formation of a Density Dip

First we will take a look at the generation of a double layer in a current carrying plasma. In 1968 Alfvén and Carlqvist showed that a density dip in a current carrying plasma can be favourable for the generation of a double layer. In this case we look at the plasma as a combination of two fluids, the moving electron fluid and the immobile ion fluid which acts as a neutralizing background. The electron fluid is treated as an essentially zero temperature beam and the ions are assumed to be collisional, and possess some finite temperature.

The density dip in the plasma (of both electrons and ions) will cause an electric field to be generated in order to keep the current density at the same level, i.e. electrons are accelerated in the decreasing part into the dip and decelerated in the increasing part out of the dip. However, this electric field will also have an influence on the first as immobile assumed ions. These ions will be driven out of the density dip, increasing it, and thereby increasing the electric field. When all ions are gone, the electric field has reached its maximum value over the dip. Note that we then have a double-double layer (increasing and decreasing electric field), and one side needs to be transported away.

We will use the quasi-static, non-relativistic description of this mechanism, which is governed by the continuity equation and the momentum equation:

$$\frac{\partial}{\partial x}(n_e v_e) = 0$$

$$m_e v_e > \frac{\partial v_e}{\partial x} = -eE$$

Combining these two equations we get an expression for the electric field dependent on the electron density:

$$E = -\frac{\partial}{\partial x}(\frac{m_e j_e^2}{2n_e^2 e^3})$$

where $j_e = -n_e e v_e$ the electron current density. The ions will experience an outward force due to this electric field,

with

$$F = en_i E = -\frac{\partial}{\partial x}(m_e n_e v2_e^2).$$

Only when the force of the electric field can overcome the force by the ion pressuregradient can the evacuation of the density dip take place. Comparing the two forces (pressure and electric) assuming a quasi-neutral, thermal plasma shows after integration that this can only happen when $k_B T_i < m_e v_e^2$. This happens to be the Bohm criterion for the stability of a double layer.

Current Carrying Double Layers Formed by Single and Zero Temperature Beams

We consider how a single zero-temperature beam of ions and a single zero-temperature beam of electrons, together with a trapped, zero velocity ion component, and a trapped, zero velocity electron component, may form a particular class of double layer. The trapped components are referred to as the 'ambient plasma' and will later be allowed to have finite temperature.

We use Poisson's equation and the conservation of momentum and number density to analyse the structure of these double layers, in the 1D, time-independent limit. We are looking for double layer like solutions, where there is a well localised region with a potential gradient, outside of which the electric field is zero. The region can be divided into the interval inside the double layer, where there is only one ion component and one electron component, but there is a finite field, and the outside region, where the electric field is zero. For the moment, we need only consider the inside region and the densities and velocities associated with the beams inside the layer.

The electron beam component is streaming with positive velocity $v_e(x)$ (to the right), and the ion beam is streaming with negative velocity $v_i(x)$ (to the left). Here, the conservation of particle energy means that $v_\alpha^2(x)/2m + q\alpha\phi(x)$ is a constant, and the conservation of particle number means that the current $j_\alpha = qn_\alpha(x)v_\alpha(x)$ is also a constant,

$$|v_\alpha(\phi_\alpha)| = \sqrt{v_{\alpha,0}^2 + \frac{2e\phi_\alpha}{m\alpha}},$$

$$n\alpha(\phi\alpha) = n_0 v\alpha_{,0}\, v_\alpha^{-1}(\phi_\alpha)$$

Where $\phi_e = \phi$ and $\phi_i = \phi_{DL} - \phi$. Here no and $v_{\alpha,0}$ are respectively the electron (and ion) density and particle drift speed at the low (high) potential side of the double layer.

Now we use Poisson's equation to obtain the maximal current through the double layer, as a function of the potential drop, the fraction of current carried by the ions as compared to the electrons and a temperature limit for the *ambient plasma*. We chose $\phi(x = 0) = 0$ and $\phi(x = d) = \phi_{DL}$, with d the thickness of the double layer.

$$\rho_e = \frac{j_e}{v_e}, \rho i = \frac{j_i}{v_i}.$$

Thus we can write Poisson's equation in the region inside the double layer as:

$$-\frac{1}{4\pi}\frac{\partial^2\phi}{\partial x^2} = \frac{j_i}{\sqrt{v^2_{i,0}+2e(\phi_{DL}-\phi)/m_i}} - \frac{j_e}{\sqrt{v^2_{e,0}+2e\phi/m_e}}.$$

Introducing an integration factor dφ/dx at both sides and integrating over x at the left hand side and over φ on the right hand side the first integration leads to the square of the electric field (dφ/dx)². The assumption that there is no electric field outside the double layer then leads to the "Langmuir condition" for non-relativistic double layers:

$$\frac{j_e}{j_i} = \frac{m_i}{m_e}.$$

For this double layer (in a hydrogen plasma) the electron current dominates the ion current by a factor of $\sqrt{1836}$. Further integration, as done by Raadu, then leads to the *Langmuir-Child* relation:

$$jd^2 = (j_e + j_i)d^2 = (1+\sqrt{\frac{me}{mi}})\frac{C_0}{9}\sqrt{\frac{2e}{m_e}}\phi_{DL}^{1.5},$$

Where C_0 is expressed in terms of the elliptical integrals E and K:

$$C_0 = 2^{-1.5}(2\sqrt{2} - E(sin\pi/8) - (1+2\sqrt{2} - K(sin\pi/8))^2 \approx 1.86518.$$

If we now allow the ambient plasma to be at finite temperature we have to take into account reflected particles more carefully and examine how far they can penetrate into the repulsive electric field. We describe the ambient plasma by a Boltzmann distribution over the double layer:

$$n_\alpha, R = n_\alpha, 0e^{-\frac{e\phi_\alpha}{k_BT\alpha}}$$

The densities of the reflected particles are now added to Poisson's equation. In order that the particles in the 'ambient plasma' be truly trapped we require that their temperature be lower than the double layer potential. This can be seen in terms of the restriction that the potential and the electric field have to vanish at the boundaries of the double layer. The precise condition is known as the "Bohm criterion":

$$m_e v_e^2 > K_B T_i.$$

A double layer of this type cannot form if this criterion is not met. This is the same condition under which a double layer can be formed by an ion density dip (or equivalently, for instability to parallel wavemodes like the ion acoustic or Buneman instability).

Vlasov-poisson Equation

In general the plasma distributions near a double layer are necessarily strongly non-Maxwellian, and therefore inaccessible to fluid models. In order to analyse double layers in full generality, the plasma must be described using the particle distribution function $f_\alpha(\vec{x},t;\vec{v})$, which describes the number of particles of species α having approximately the velocity $\vec{v}$ near the place $\vec{x}$ and time t.

The Vlasov-Poisson equations give the time-dependent interaction of a plasma (described using the particle distribution) with its self-consistent electric field. The Vlasov-Poisson equations are a combination of the Vlasov equations for each species α (also called the collisionless Boltzmann equation (CBE). We take the nonrelativistic zero-magnetic field limit):

$$\frac{\partial}{\partial t} f_\alpha + \vec{v} \frac{\partial}{\partial \vec{x}} f_\alpha + \frac{q_\alpha \vec{E}}{m_\alpha} \cdot \frac{\partial}{\partial v} f_\alpha = 0,$$

and Poisson's equation:

$$\nabla \cdot \vec{E} = -\frac{\partial 2\phi}{\partial x^2} = 4\pi\rho.$$

Here q_α is the particle's electric charge, m_α is the particle's mass, $\vec{E}(\vec{x},t)$ is the electric field, $\phi(\vec{x},t)$ the electric potential and ρ is the electric charge density.

THOMSON SCATTERING

Thomson scattering is the elastic scattering of electromagnetic radiation by a free charged particle, as described by classical electromagnetism. It is just the low-energy limit of Compton scattering: the particle's kinetic energy and photon frequency do not change as a result of the scattering. This limit is valid as long as the photon energy is much smaller than the mass energy of the particle: $v \ll mc^2 / h$, or equivalently, if the wavelength of the light is much greater than the Compton wavelength of the particle.

Description of the Phenomenon

In the low-energy limit, the electric field of the incident wave (photon) accelerates the charged particle, causing it, in turn, to emit radiation at the same frequency as the incident wave, and thus the wave is scattered. Thomson scattering is an important phenomenon in plasma physics and was first explained by the physicist J. J. Thomson. As long as the motion of the particle is non-relativistic (i.e. its speed is much less than the speed of light), the main cause of the acceleration of the particle will be due to the electric field component of the incident wave. In a first approximation, the influence of the magnetic field can be neglected. The particle will move in the direction of the oscillating electric field, resulting in electromagnetic dipole radiation. The moving particle radiates most strongly in a direction perpendicular to its acceleration and that radiation will be polarized along the direction of its motion. Therefore, depending on where an observer is located, the light scattered from a small volume element may appear to be more or less polarized.

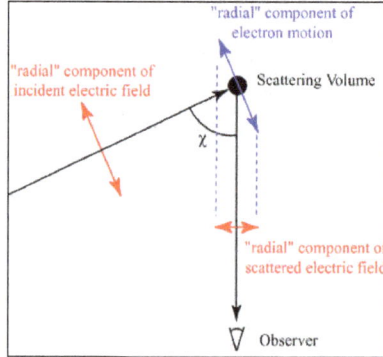

The electric fields of the incoming and observed wave (i.e. the outgoing wave) can be divided up into those components lying in the plane of observation (formed by the incoming and observed waves) and those components perpendicular to that plane. Those components lying in the plane are referred to as "radial" and those perpendicular to the plane are "tangential". (It is difficult to make these terms seem natural, but it is standard terminology.)

The diagram on the right depicts the plane of observation. It shows the radial component of the incident electric field, which causes the charged particles at the scattering point to exhibit a radial component of acceleration (i.e., a component tangent to the plane of observation). It can be shown that the amplitude of the observed wave will be proportional to the cosine of χ, the angle between the incident and observed waves. The intensity, which is the square of the amplitude, will then be diminished by a factor of $\cos^2(\chi)$. It can be seen that the tangential components (perpendicular to the plane of the diagram) will not be affected in this way.

The scattering is best described by an emission coefficient which is defined as ε where ε dt dV dΩ dλ is the energy scattered by a volume element dV in time dt into solid angle dΩ between wavelengths λ and λ+dλ. From the point of view of an observer, there are two emission coefficients, ε_r corresponding to radially polarized light and ε_t corresponding to tangentially polarized light. For unpolarized incident light, these are given by:

$$\varepsilon_t = \frac{\pi\sigma_t}{2} In$$

$$\varepsilon_r = \frac{\pi\sigma_t}{2} In\cos^2\chi$$

where n is the density of charged particles at the scattering point, I is incident flux (i.e. energy/time/area/wavelength) and σ_t is the Thomson cross section for the charged particle, defined below. The total energy radiated by a volume element dV in time dt between wavelengths λ and $\lambda+d\lambda$ is found by integrating the sum of the emission coefficients over all directions (solid angle):

$$\int \varepsilon \, d\Omega = \int_0^{2\pi} d\varphi \int_0^{\pi} d\chi (\varepsilon_t + \varepsilon_r) \sin \chi = I\sigma_t n (2 + 2/3)\pi^2 = I\sigma_t n \frac{8}{3}\pi^2.$$

The Thomson differential cross section, related to the sum of the emissivity coefficients, is given by,

$$\frac{d\sigma_t}{d\Omega} = \left(\frac{q^2}{4\pi\varepsilon_0 mc^2} \right)^2 \frac{1+\cos^2 \chi}{2}$$

expressed in SI units; q is the charge per particle, m the mass of particle, and a constant, the permittivity of free space. (To obtain an expression in cgs units, drop the factor of $4\pi\varepsilon_0$.) Integrating over the solid angle, we obtain the Thomson cross section

$$\sigma_t = \frac{8\pi}{3} \left(\frac{q^2}{4\pi\varepsilon_0 mc^2} \right)^2 \text{ in SI units.}$$

The important feature is that the cross section is independent of photon frequency. The cross section is proportional by a simple numerical factor to the square of the classical radius of a point particle of mass m and charge q, namely,

$$\sigma_t = \frac{8\pi}{3} r_e^2$$

Alternatively, this can be expressed in terms of λ_c the Compton wavelength, and the fine structure constant:

$$\sigma_t = \frac{8\pi}{3} \left(\frac{\alpha\lambda_c}{2\pi} \right)^2$$

For an electron, the Thomson cross-section is numerically given by:

$$\sigma_t = \frac{8\pi}{3} \left(\frac{\alpha\hbar c}{mc^2} \right)^2 = 6.6524587158\ldots \times 10^{-29} \text{ m}^2 = 66.524587158\ldots \text{ (fm)}^2$$

Examples of Thomson Scattering

The cosmic microwave background is linearly polarized as a result of Thomson scattering, as measured by DASI and more recent experiments.

The solar K-corona is the result of the Thomson scattering of solar radiation from solar coronal electrons. The ESA and NASA SOHO mission and the NASA STEREO mission generate three-dimensional images of the electron density around the sun by measuring this K-corona from three separate satellites.

In tokamaks, corona of ICF targets and other experimental fusion devices, the electron temperatures and densities in the plasma can be measured with high accuracy by detecting the effect of Thomson scattering of a high-intensity laser beam.

Inverse-Compton scattering can be viewed as Thomson scattering in the rest frame of the relativistic particle.

X-ray crystallography is based on Thomson scattering.

PLASMA STABILITY

One on the top of a hill (left) will accelerate away from its rest point if perturbed, and is thus dynamically unstable. Plasmas have many mechanisms that make them fall into the second group under certain conditions.

A ball at rest in a valley (right) will return to the bottom
if moved slightly, or perturbed, and is thus dynamically stable.

The stability of a plasma is an important consideration in the study of plasma physics. When a system containing a plasma is at equilibrium, it is possible for certain parts of the plasma to be disturbed by small perturbative forces acting on it. The stability of the system determines if the perturbations will grow, oscillate, or be damped out.

In many cases, a plasma can be treated as a fluid and its stability analyzed with magnetohydrodynamics (MHD). MHD theory is the simplest representation of a plasma, so MHD stability is a necessity for stable devices to be used for nuclear fusion, specifically magnetic fusion energy. There are, however, other types of instabilities, such as velocity-space instabilities in magnetic mirrors and systems with beams. There are also rare cases of systems, e.g. the field-reversed configuration, predicted by MHD to be unstable, but which are observed to be stable, probably due to kinetic effects.

Plasma Instabilities

Plasma instabilities can be divided into two general groups:

- Hydrodynamic instabilities.

- Kinetic instabilities.

Plasma instabilities are also categorised into different modes (e.g. with reference to a particle beam):

Mode (azimuthal wave number)	Note	Description	Radial modes	Description
m=0		Sausage instability: displays harmonic variations of beam radius with distance along the beam axis.	n=0	Axial hollowing
			n=1	Standard sausaging
			n=2	Axial bunching
m=1		Sinuous, kink or hose ins-tability: represents transverse displacements of the beam cross-section without change in the form or in a beam characteristics other than the position of its center of mass.		
m=2	Filamentation modes: growth leads towards the breakup of the beam into separate filaments.	Gives an elliptic cross-section.		
m=3		Gives a pyriform (pear-shaped) cross-section.		
m=4		Consists of four intertwined helices.		

MHD Instabilities

Beta is a ratio of the plasma pressure over the magnetic field strength,

$$\beta = \frac{p}{p_{mag}} = \frac{nk_B T}{(B^2 / 2\mu_0)}$$

MHD stability at high beta is crucial for a compact, cost-effective magnetic fusion reactor. Fusion power density varies roughly as β^2 at constant magnetic field, or as β_N^4 at constant bootstrap fraction in configurations with externally driven plasma current. (Here $\beta_N = \beta / (I / aB)$ is the normalized beta.) In many cases MHD stability represents the primary limitation on beta and thus on fusion power density. MHD stability is also closely tied to issues of creation and sustainment of certain magnetic configurations, energy confinement, and steady-state operation. Critical issues include understanding and extending the stability limits through the use of a variety of plasma configurations, and developing active means for reliable operation near those limits. Accurate predictive capabilities are needed, which will require the addition of new physics to existing MHD models. Although a wide range of magnetic configurations exist, the underlying MHD physics is common to all. Understanding of MHD stability gained in one configuration can benefit others, by verifying analytic theories, providing benchmarks for predictive MHD stability codes, and advancing the development of active control techniques.

The most fundamental and critical stability issue for magnetic fusion is simply that MHD instabilities often limit performance at high beta. In most cases the important instabilities are long wavelength, global modes, because of their ability to cause severe degradation of energy confinement or termination of the plasma. Some important examples that are common to many magnetic configurations are ideal kink modes, resistive wall modes, and neoclassical tearing modes. A possible consequence of violating stability boundaries is a disruption, a sudden loss of thermal energy often followed by termination of the discharge. The key issue thus includes understanding the nature of the beta limit in the various configurations, including the associated thermal and magnetic stresses, and finding ways to avoid the limits or mitigate the consequences. A wide range of approaches to preventing such instabilities is under investigation, including optimization of the configuration of the plasma and its confinement device, control of the internal structure of the plasma, and active control of the MHD instabilities.

Ideal Instabilities

Ideal MHD instabilities driven by current or pressure gradients represent the ultimate operational limit for most configurations. The long-wavelength kink mode and short-wavelength ballooning mode limits are generally well understood and can in principle be avoided.

Intermediate-wavelength modes (n ~ 5–10 modes encountered in tokamak edge plasmas, for example) are less well understood due to the computationally intensive nature of the stability calculations. The extensive beta limit database for tokamaks is consistent with ideal MHD stability limits, yielding agreement to within about 10% in beta for cases where the internal profiles of the plasma are accurately measured. This good agreement provides confidence in ideal stability calculations for other configurations and in the design of prototype fusion reactors.

Resistive Wall Modes

Resistive wall modes (RWM) develop in plasmas that require the presence of a perfectly conducting wall for stability. RWM stability is a key issue for many magnetic configurations. Moderate beta values are possible without a nearby wall in the tokamak, stellarator, and other configurations, but a nearby conducting wall can significantly improve ideal kink mode stability in most configurations, including the tokamak, ST, reversed field pinch (RFP), spheromak, and possibly the FRC. In the advanced tokamak and ST, wall stabilization is critical for operation with a large bootstrap fraction. The spheromak requires wall stabilization to avoid the low-m, n tilt and shift modes, and possibly bending modes. However, in the presence of a non-ideal wall, the slowly growing RWM is unstable. The resistive wall mode has been a long-standing issue for the RFP, and has more recently been observed in tokamak experiments. Progress in understanding the physics of the RWM and developing the means to stabilize it could be directly applicable to all magnetic configurations. A closely related issue is to understand plasma rotation, its sources and sinks, and its role in stabilizing the RWM.

Resistive Instabilities

Resistive instabilities are an issue for all magnetic configurations, since the onset can occur at beta values well below the ideal limit. The stability of neoclassical tearing modes (NTM) is a key issue for magnetic configurations with a strong bootstrap current. The NTM is a metastable mode; in certain plasma configurations, a sufficiently large deformation of the bootstrap current produced by a "seed island" can contribute to the growth of the island. The NTM is already an important performance-limiting factor in many tokamak experiments, leading to degraded confinement or disruption. Although the basic mechanism is well established, the capability to predict the onset in present and future devices requires better understanding of the damping mechanisms which determine the threshold island size, and of the mode coupling by which other instabilities (such as sawteeth in tokamaks) can generate seed islands. Resistive Ballooning Mode, similar to ideal ballooning, but with finite resistivity taken into consideration, provides another example of a resistive instability.

Opportunities for Improving MHD Stability

Configuration

The configuration of the plasma and its confinement device represent an opportunity to improve MHD stability in a robust way. The benefits of discharge shaping and low aspect ratio for ideal MHD stability have been clearly demonstrated in tokamaks and STs, and will continue to be investigated in experiments such as DIII-D, Alcator C-Mod, NSTX, and MAST. New stellarator experiments such as NCSX (proposed) will test the prediction that addition of appropriately designed helical coils can stabilize ideal kink modes at high beta, and lower-beta tests of ballooning stability are possible in HSX. The new ST experiments provide an opportunity to test predictions that a low aspect ratio yields improved stability to tearing modes, including neoclassical, through a large stabilizing "Glasser effect" term associated with a large Pfirsch-Schlüter current. Neoclassical tearing modes can be avoided by minimizing the bootstrap current in quasi-helical and quasi-omnigenous stellarator configurations. Neoclassical tearing modes are also stabilized with the appropriate relative signs of the bootstrap current and the magnetic shear; this prediction is supported by the absence of NTMs in central negative shear regions of tokamaks. Stellarator configurations such as the proposed NCSX, a quasi-axisymmetric stellarator design, can be created with negative magnetic shear and positive bootstrap current to achieve stability to the NTM. Kink mode stabilization by a resistive wall has been demonstrated in RFPs and tokamaks, and will be investigated in other configurations including STs (NSTX) and spheromaks (SSPX). A new proposal to stabilize resistive wall modes by a flowing liquid lithium wall needs further evaluation.

Internal Structure

Control of the internal structure of the plasma allows more active avoidance of MHD instabilities. Maintaining the proper current density profile, for example, can help to maintain stability to tearing modes. Open-loop optimization of the pressure and current density profiles with external heating and current drive sources is routinely used in many devices. Improved diagnostic measurements along with localized heating and current drive sources, now becoming available, will allow active feedback control of the internal profiles in the near future. Such work is beginning or planned in most of the large tokamaks (JET, JT–60U, DIII–D, C–Mod, and ASDEX–U) using RF heating and current drive. Real-time analysis of profile data such as MSE current profile measurements and real-time identification of stability boundaries are essential components of profile control. Strong plasma rotation can stabilize resistive wall modes, as demonstrated in tokamak experiments, and rotational shear is also predicted to stabilize resistive modes. Opportunities to test these predictions are provided by configurations such as the ST, spheromak, and FRC, which have a large natural diamagnetic rotation, as well as tokamaks with rotation driven by neutral beam injection. The Electric Tokamak experiment is intended to have a very large driven rotation, approaching Alfvénic regimes where ideal stability may also be influenced. Maintaining sufficient plasma rotation, and the possible role of the RWM in damping the rotation, are important issues that can be investigated in these experiments.

Feedback Control

Active feedback control of MHD instabilities should allow operation beyond the "passive" stability limits. Localized RF current drive at the rational surface is predicted to reduce or eliminate neoclassical tearing mode islands. Experiments have begun in ASDEX–U and COMPASS-D with promising results, and are planned for next year in DIII–D. Routine use of such a technique in generalized plasma conditions will require real-time identification of the unstable mode and its radial location. If the plasma rotation needed to stabilize the resistive wall mode cannot be maintained, feedback stabilization with external coils will be required. Feedback experiments have begun in DIII–D and HBT-EP, and feedback control should be explored for the RFP and other configurations. Physics understanding of these active control techniques will be directly applicable between configurations.

Disruption Mitigation

The techniques for improving MHD stability are the principal means of avoiding disruptions. However, in the event that these techniques do not prevent an instability, the effects of a disruption can be mitigated by various techniques. Experiments in JT–60U have demonstrated reduction of electromagnetic stresses through operation at a neutral point for vertical stability. Pre-emptive removal of the plasma energy by injection of a large gas puff or an impurity pellet has been demonstrated in tokamak experiments, and ongoing experiments in C–Mod, JT–60U, ASDEX–U, and DIII–D will improve the understanding and predictive capability. Cryogenic liquid jets of helium are another proposed technique, which may be required for larger devices. Mitigation techniques developed for tokamaks will be directly applicable to other configurations.

Farley–Buneman Instability

The Farley–Buneman instability, or FB instability, is a microscopic plasma instability named after Donald T. Farley and Oscar Buneman. It is similar to the ionospheric Rayleigh-Taylor instability.

It occurs in collisional plasma with neutral component, and is driven by drift currents. It can be thought of as a modified two-stream instability arising from the difference in drifts of electrons and ions exceeding the ion acoustic speed.

It is present in the equatorial and polar ionospheric E-regions. In particular, it occurs in the equatorial electrojet due to the drift of electrons relative to ions, and also in the trails behind ablating meteoroids. Since the FB fluctuations can scatter electromagnetic waves, the instability can be used to diagnose the state of ionosphere by the use of electromagnetic pulses.

Conditions

To derive the dispersion relation below, we make the following assumptions. First, quasi-neutrality is assumed. This is appropriate if we restrict ourselves to wavelengths longer than the Debye length. Second, the collision frequency between ions and background neutral particles is assumed to be much greater than the ion cyclotron frequency, allowing the ions to be treated as unmagnetized. Third, the collision frequency between electrons and background neutrals is assumed to be much less than the electron cyclotron frequency. Finally, we only analyze low frequency waves so that we can neglect electron inertia. Because the Buneman instability is electrostatic in nature, only electrostatic perturbations are considered.

Dispersion Relation

We use linearized fluid equations (equation of motion, equation of continuity) for electrons and ions with Lorentz force and collisional terms. The equation of motion for each species is:

Electrons: $0 = -en(\vec{E} + \vec{v}_e \times \vec{B}) - k_b T_e \nabla n - m_e n v_{en} \vec{v}_e$

Ions: $m_i n \dfrac{dv_i}{dt} = en(\vec{E} + \vec{v}_i \times \vec{B}) - k_b T_i \nabla n - m_i n v_{in} \vec{v}_i$

where:

- m_s is the mass of species.

- v_s is the velocity of species.

- T_s is the temperature of species.

- v_{sn} is the frequency of collisions between species s and neutral particles.

- e is the charge of an electron.

- n is the electron number density.

- k_b is the Boltzmann Constant.

Note that electron inertia has been neglected, and that both species are assumed to have the same number density at every point in space ($n_i = n_e = n$). The collisional term describes the momentum loss frequency of each fluid due to collisions of charged particles with neutral particles in the plasma. We denote v_{en} as the frequency of collisions between electrons and neutrals, and as the frequency of collisions between ions and neutrals. We also assume that all perturbed properties, such as species velocity, density, and the electric field, behave as plane waves. In other words, all physical quantities f will behave as an exponential function of time t and position x (where k is the wave number):

$f \sim \exp(i\omega t + ikx)$.

This can lead to oscillations if the frequency ω is a real number, or to either exponential growth or exponential decay if ω is complex. If we assume that the ambient electric and magnetic fields are perpendicular to one another and only analyze waves propagating perpendicular to both of these fields, the dispersion relation takes the form of:

$$\omega\left(1 + i\psi_0 \frac{\omega - iv_{in}}{v_{in}}\right) = kv_E + i\psi_0 \frac{k^2 c_i^2}{v_{in}},$$

where v_E is the E$\times$B drift and c_i is the acoustic speed of ions. The coefficient ψ_0 described the combined effect of electron and ion collisions as well as their cyclotron frequencies Ω_i and Ω_e.

Growth Rate

Solving the dispersion we arrive at frequency given as:

$$\omega = \omega_r + i\gamma,$$

Where γ describes the growth rate of the instability. For FB we have the following:

$$\omega_r = \frac{kv_E}{1+\psi_0}$$

$$\gamma = \frac{\psi_0}{v_{in}} \frac{\omega_r^2 - k^2 c_i^2}{1+\psi_0}.$$

Interchange Instability

The interchange instability is a type of plasma instability seen in magnetic fusion energy that is driven by the gradients in the magnetic pressure in areas where the confining magnetic field is curved. The name of the instability refers to the action of the plasma changing position with the magnetic field lines (i.e. an interchange of the lines of force in space) without significant disturbance to the geometry of the external field. The instability causes flute-like structures to appear on the surface of the plasma, and thus the instability is also known as the flute instability. The interchange instability is a key issue in the field of fusion energy, where magnetic fields are used to confine a plasma in a volume surrounded by the field.

The basic concept was first noted in a famous 1954 paper by Martin David Kruskal and Martin Schwarzschild, which demonstrated that a situation similar to the Rayleigh–Taylor instability in classic fluids existed in magnetically confined plasmas. The problem can occur anywhere where the magnetic field is concave with the plasma on the inside of the curve. Edward Teller gave a talk on the issue at a meeting later that year, pointing out that it appeared to be an issue in most of the fusion devices being studied at that time. He used the analogy of rubber bands on the outside of a blob of jelly; there is a natural tendency for the bands to snap together and eject the jelly from the center.

Most machines of that era were suffering from other instabilities that were far more powerful, and whether or not the interchange instability was taking place could not be confirmed. This was finally demonstrated beyond doubt by a Soviet magnetic mirror machine during an international meeting in 1961. When the US delegation stated they were not seeing this problem in their mirrors, it was pointed out they were making an error in the use of their instrumentation. When that was considered, it was clear the US experiments were also being affected by the same problem. This led to a series of new mirror designs, as well as modifications to other designs like the stellarator to add negative curvature. These had cusp-shaped fields so that the plasma was contained within convex fields, the so-called "magnetic well" configuration.

In modern designs, the interchange instability is suppressed by the complex shaping of the fields. In the tokamak design there are still areas of "bad curvature", but particles within the plasma spend only a short time in those areas before being circulated to an area of "good curvature". Modern stellarators use similar configurations, differing from tokamaks largely in how that shaping is created.

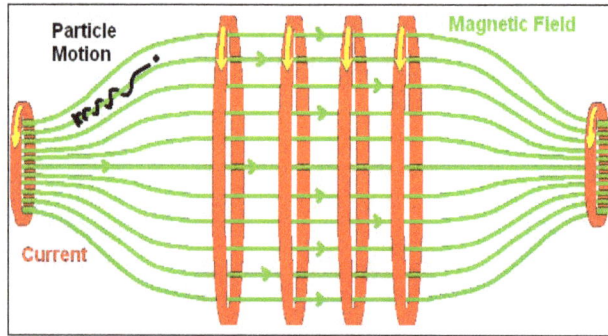

A basic magnetic mirror.

The magnetic lines of force (*green*) confine plasma particles by causing them to rotate around the lines (*black*). As the particles approach the ends of the mirror, they see an increasing force back into the center of the chamber. Ideally, all particles would continue to be reflected and stay within the machine.

Most magnetic confinement systems try to hold the plasma within a vacuum chamber using magnetic fields. The plasma particles are electrically charged, and thus see a traverse force from the field. When the particle's original linear motion is superimposed on this traverse force, its resulting path through space is a helix, or corkscrew shape. Since the electrons are much lighter than the ions, they move in a tighter orbit. Such a field will thus trap the plasma by forcing it to flow along the lines. Properly arranged, a magnetic field can prevent the plasma from reaching the outside of the field where they would impact with the vacuum chamber. The fields should also try to keep the ions and electrons mixed - so charge separation does not occur.

The magnetic mirror is one example of a simple magnetic plasma trap. The mirror has a field that runs along the open center of the cylinder and bundles together at the ends. In the center of the chamber the particles follow the lines and flow towards either end of the device. There, the increasing magnetic density causes them to "reflect", reversing direction and flowing back into the center again. Ideally, this will keep the plasma confined indefinitely, but even in theory there a critical angle between the particle trajectory and the axis of the mirror where particles can escape. Initial calculations showed that the loss rate through this process would be small enough to not be a concern. However, in practice, all mirror machines demonstrated a loss rate far higher than these calculations suggested.

The interchange instability was one of the major reasons for these losses. The mirror field has a cigar shape to it, with increasing curvature at the ends. When the plasma is located in its design location, the electrons and ions are roughly mixed. However, if the plasma is displaced, the non-uniform nature of the field means the ion's larger orbital radius takes them outside the confinement area while the electrons remain inside. It is possible the ion will hit the wall of the container, removing it from the plasma. If this occurs, the outer edge of the plasma is now net negatively charged, attracting more of the positively charged ions, which then escape as well.

This effect allows even a tiny displacement to drive the entire plasma mass to the walls of the container. The same effect occurs in any reactor design where the plasma is within a field of sufficient curvature, which includes the outside curve of toroidal machines like the tokamak and stellarator. As this process is highly non-linear, it tends to occur in isolated areas, giving rise to the flute-like expansions as opposed to mass movement of the plasma as a whole.

Instability in a Plasma System

The single most important property of a plasma is its stability. MHD and its derived equilibrium equations offer a wide variety of plasmas configurations but the stability of those configurations have not been challenged. More specifically, the system must satisfy the simple condition where ? is the change in potential energy for degrees of freedom. Failure to meet this condition indicates that there is a more energetically preferable state. The system will evolve and either shift into a different state or never reach a steady state. These instabilities pose great challenges to those aiming to make stable plasma configurations in the lab. However, they have also granted us an informative tool on the behavior of plasma, especially in the examination of planetary magnetospheres.

This process injects hotter, lower density plasma into a colder, higher density region. It is the MHD analog of the well-known Rayleigh-Taylor instability. The Rayleigh-Taylor instability occurs at an interface in which a lower density liquid pushes against a higher density liquid in a gravitational field. In a similar model with a gravitational field, the interchange instability acts in the same way. However, in planetary magnetospheres co-rotational forces are dominant and change the picture slightly.

Simple Models

Let's first consider the simple model of a plasma supported by a magnetic field B in a uniform gravitational field g. To simplify matters, assume that the internal energy of the system is zero such that static equilibrium may be obtained from the balance of the gravitational force and the magnetic field pressure on the boundary of the plasma. If two adjacent flux tubes lying opposite along the boundary (one fluid tube and one magnetic flux tube) are interchanged the volume element doesn't change and the field lines are straight. Therefore, the magnetic potential doesn't change, but the gravitational potential changes since it was moved along the z axis. Since the change in is negative the potential is decreasing. A decreasing potential indicates a more energetically favorable system and consequently an instability. The origin of this instability is in the J × B forces that occur at the boundary between the plasma and magnetic field. At this boundary there are slight ripple-like perturbations in which the low points must have a larger current than the high points since at the low point more gravity is being supported against the gravity. The difference in current allows negative and positive charge to build up along the opposite sides of the valley. The charge build-up produces an E field between the hill and the valley. The accompanying E × B drifts are in the same direction as the ripple, amplifying the effect. This is what is physically meant by the "interchange" motion.

These interchange motions also occur in plasmas that are in a system with a large centrifugal force. In a cylindrically symmetric plasma device, radial electric fields cause the plasma to rotate rapidly in a column around the axis. Acting opposite to the gravity in the simple model, the centrifugal force moves the plasma outward where the ripple-like perturbations (sometimes called "flute" instabilities) occur on the boundary. This is important for the study of the magnetosphere in which the co-rotational forces are stronger than the opposing gravity of the planet. Effectively, the less dense "bubbles" inject radially inward in this configuration. Without gravity or an inertial force, interchange instabilities can still occur if the plasma is in a curved magnetic field. If we assume the potential energy to be purely magnetic then the change in potential energy is: If the fluid is incompressible then the equation can be simplified into. Since (to maintain pressure balance), the above equation shows that if the system is unstable. Physically, this means that if the field lines are toward the region of higher plasma density then the system is susceptible to interchange motions. To derive a more rigorous stability condition, the perturbations that cause an instability must be generalized. The momentum equation for a resistive MHD is linearized and then manipulated into a linear force operator. Due to purely mathematical reasons, it is then possible to split the analysis into two approaches: the normal mode method and the energy method. The normal mode method essentially looks for the eigenmodes and eigenfrequencies and summing the solutions to form the general solution. The energy method is similar to the simpler approach outlined above where is found for any arbitrary perturbation in order to maintain the condition. These two methods are not exclusive and can be used together to establish a reliable diagnosis of the stability.

Observations in Space

The strongest evidence for interchange transport of plasma in any magnetosphere is the observation of injection events. The recording of these events in the magnetospheres of Earth, Jupiter and Saturn are the main tool for the interpretation and analysis of interchange motion.

Earth

Although spacecraft have travelled many times in the inner and outer orbit of Earth since the 1960s, the spacecraft ATS 5 was the first major plasma experiment performed that could reliably determine the existence of radial injections driven by interchange motions. The analysis revealed the frequent injection of a hot plasma cloud is injected inward during a substorm in the outer layers of the magnetosphere. The injections occur predominantly in the night-time hemisphere, being associated with the depolarization of the neutral sheet configuration in the tail regions of the magnetosphere. This implies that Earth's magnetotail region is a major mechanism in which the magnetosphere stores and releases energy through the interchange mechanism. The interchange instability also has been found to have a limiting factor on the night side plasmapause thickness., the plasmapause is found to be near the geosynchronous orbit in which the centrifugal and gravitational potential cancel exactly. This sharp change in plasma pressure associated with the plasma pause enables this instability. A mathematical treatment comparing the growth rate of the instability with the thickness of the plasmapause boundary revealed that the interchange instability limits the thickness of the boundary.

Jupiter

Interchange instability plays a major role in the radial transport of plasma in the Io plasma torus at Jupiter. The first evidence of this behavior was published by Thorne et al. in which they discovered "anomalous plasma signatures" in the Io torus of Jupiter's magnetosphere. Using the data from Galileo's energetic particle detector (EPD), the study looked at one specific event. In Thorne et al. they concluded that these events had a density differential of at least a factor of 2, a spatial scale of km and an inward velocity of about km/s. These results support the theoretical arguments for interchange transport. Later, more injections events were discovered and analyzed from Galileo. Mauk et al. used over 100 Jovian injections to study how these events were dispersed in energy and time. Similar to injections of Earth, the events were often clustered in time. The authors concluded that this indicated the injection events were triggered by solar wind activity against the Jovian magnetosphere. This is very similar to the magnetic storm relationship injection events have on Earth. However, it was found that Jovian injections can occur at all local time positions and therefore can't be directly related to the situation in Earth's magnetosphere. Although the Jovian injections aren't a direct analog of Earth's injections, the similarities indicate that this process plays a vital role in the storage and release of energy. The difference may lie in the presence of Io in the Jovian system. Io is a large producer of plasma mass because of its volcanic activity. This explains why the bulk of interchange motions are seen in a small radial range near Io.

Saturn

Recent evidence from the spacecraft Cassini has confirmed that the same interchange process is prominent on Saturn. Unlike Jupiter, the events happen much more frequently and more clearly. The difference lies in the configuration of the magnetosphere. Since Saturn's gravity is much weaker, the gradient/curvature drift for a given particle energy and L value is about 25 times faster. Saturn's magnetosphere provides a much better environment for the study of interchange instability under these conditions even though the process is essential in both Jupiter and Saturn. In a case study of one injection event, the Cassini Plasma Spectrometer (CAPS) produced characteristic radial profiles of plasma densities and temperatures of the plasma particles that also allowed the calculation of the origin of the injection and the radial propagation velocity. The electron density inside the event was lowered by a factor of about 3, the electron temperature was higher by an order of magnitude than the background, and there was a slight increase in the magnetic field. The study also used a model of pitch angle distributions to estimate the event originated between $9 < L < 11$ and had a radial speed of about 260+60/-70 km/s.. The similarities imply that the Saturn and Jupiter processes are the same.

Electrothermal Instability

The electrothermal instability (also known as ionization instability, non-equilibrium instability or Velikhov instability in the literature) is a magnetohydrodynamic (MHD) instability appearing in magnetized non-thermal plasmas used in MHD converters. It was first theoretically discovered in 1962 and experimentally measured into a MHD generator in 1963 by Evgeny Velikhov.

This instability is a turbulence of the electron gas in a non-equilibrium plasma (i.e. where the electron temperature T_e is greatly higher than the overall gas temperature T_g). It arises when a magnetic field powerful enough is applied in such a plasma, reaching a critical Hall parameter β_{cr}.

Locally, the number of electrons and their temperature fluctuate (electron density and thermal velocity) as the electric current and the electric field.

The Velikhov instability is a kind of ionization wave system, almost frozen in the two temperature gas. The reader can evidence such a stationary wave phenomenon just applying a transverse magnetic field with a permanent magnet on the low-pressure control gauge (Geissler tube) provided on vacuum pumps. In this little gas-discharge bulb a high voltage electric potential is applied between two electrodes which generates an electric glow discharge (pinkish for air) when the pressure has become low enough. When the transverse magnetic field is applied on the bulb, some oblique grooves appear in the plasma, typical of the electrothermal instability.

The electrothermal instability occurs extremely quickly, in a few microseconds. The plasma becomes non-homogeneous, transformed into alternating layers of high free electron and poor free electron densities. Visually the plasma appears stratified, as a "pile of plates".

Hall Effect in Plasmas

The Hall effect in ionized gases has nothing to do with the Hall effect in solids (where the Hall parameter is always very inferior to unity). In a plasma, the Hall parameter can take any value.

The Hall parameter β in a plasma is the ratio between the electron gyrofrequency Ω_e and the electron-heavy particles collision frequency v:

$$\beta = \frac{\Omega_e}{v} = \frac{e\,B}{m_e\,v}$$

where:

> e is the electron charge (1.6×10^{-19} coulomb).
>
> B is the magnetic field (in teslas).
>
> m_e is the electron mass (0.9×10^{-30} kg).

The Hall parameter value increases with the magnetic field strength.

Physically, when the Hall parameter is low, the trajectories of electrons between two encounters with heavy particles (neutral or ion) are almost linear. But if the Hall parameter is high, the electron movements are highly curved. The current density vector J is no more colinear with the electric field vector E. The two vectors J and E make the Hall angle θ which also gives the Hall parameter:

$$\beta = \tan\theta$$

Plasma Conductivity and Magnetic Fields

In a non-equilibrium ionized gas with high Hall parameter, Ohm's law,

$$\mathbf{J} = \sigma\mathbf{E}$$

where σ is the electrical conductivity (in siemens per metre),

is a matrix, because the electrical conductivity σ is a matrix:

$$\sigma = \sigma_s \left\| \begin{matrix} \dfrac{1}{1+\beta^2} & \dfrac{-\beta}{1+\beta^2} \\[2mm] \dfrac{\beta}{1+\beta^2} & \dfrac{1}{1+\beta^2} \end{matrix} \right\|$$

σ_s is the scalar electrical conductivity:

$$\sigma_s = \frac{n_e\,e^2}{m_e\,v}$$

where n_e is the electron density (number of electrons per cubic meter).

The current density J has two components:

$$J_{\parallel} = \frac{n_e \, e^2}{m_e \, v} \frac{E}{1+\beta^2} \quad \text{and} \quad J_{\perp} = \frac{-n_e \, e^2}{m_e \, v} \frac{\beta \, E}{1+\beta^2}$$

Therefore,

$$J_{\perp} = J_{\parallel} \, \beta$$

The Hall effect makes electrons "crabwalk".

When the magnetic field B is high, the Hall parameter β is also high, and $\frac{1}{1+\beta^2} \ll 1$. Thus both conductivities,

$$\sigma_{\parallel} \approx \frac{\sigma_s}{\beta^2} \quad \text{and} \quad \sigma_{\perp} \approx \frac{\sigma_s}{\beta}$$

become weak, therefore the electric current cannot flow in these areas. This explains why the electron current density is weak where the magnetic field is the strongest.

Critical Hall Parameter

The electrothermal instability occurs in a plasma at a ($T_e > T_g$) regime when the Hall parameter is higher that a critical value β_{cr}.

We have,

$$f = \frac{\left(\dfrac{\delta\mu}{\mu}\right)}{\left(\dfrac{\delta n_e}{n_e}\right)}$$

where μ is the electron mobility (in m²/(V·s)) and

$$s = \frac{2 \, k \, T_e^2}{E_i \, (T_e - T_g)} \times \frac{1}{1+\dfrac{3}{2}\dfrac{k \, T_e}{E_i}}$$

where E_i is the ionization energy (in electron volts) and k the Boltzmann constant.

The growth rate of the instability is,

$$g = \frac{\sigma \, E^2}{n_e \left(E_i + \dfrac{3}{2} k \, T_e \right)\left(1+\beta^2\right)} (\beta - \beta_{cr})$$

And the critical Hall parameter is:

$$\beta_{cr} = 1.935f + 0.065 + s$$

The critical Hall parameter β_{cr} greatly varies according to the degree of ionization α:

$$\alpha = \frac{n_i}{n_n}$$

where n_i is the ion density and n_n the neutral density (in particles per cubic metre).

The electron-ion collision frequency ν_{ei} is much greater than the electron-neutral collision frequency ν_{en}.

Therefore, with a weak energy degree of ionization α, the electron-ion collision frequency ν_{ei} can equal the electron-neutral collision frequency ν_{en}.

- For a weakly ionized gas (non-Coulombian plasma, when $\nu_{ei} < \nu_{en}$):

$$\beta_{cr} \approx (s^2 + 2s)^{\frac{1}{2}}$$

- For a fully ionized gas (Coulombian plasma, when $\nu_{ei} > \nu_{en}$):

$$\beta_{cr} \approx (2 + s)$$

NB: The term "fully ionized gas", introduced by Lyman Spitzer, does not mean the degree of ionization is unity, but only that the plasma is Coulomb-collision dominated, which can correspond to a degree of ionization as low as 0.01%.

Technical Problems and Solutions

A two-temperature gas, globally cool but with hot electrons ($T_e \gg T_g$) is a key feature for practical MHD converters, because it allows the gas to reach sufficient electrical conductivity while protecting materials from thermal ablation. This idea was first introduced for MHD generators in the early 1960s by Jack L. Kerrebrock and Alexander E. Sheindlin.

But the unexpected large and quick drop of current density due to the electrothermal instability ruined many MHD projects worldwide, while previous calculation envisaged energy conversion efficiencies over 60% with these devices. Whereas some studies were made about the instability by various researchers, no real solution was found at that time. This prevented further developments of non-equilibrium MHD generators and compelled most engaged countries to cancel their MHD power plants programs and to retire completely from this research field in the early 1970s, because this technical problem was considered as an impassable stumbling block in these days.

Nevertheless, experimental studies about the growth rate of the electrothermal instability and the critical conditions showed that a stability region still exists for high electron temperatures. The stability is given by a quick transition to "fully ionized" conditions (fast enough to overtake the growth rate of the electrothermal instability) where the Hall parameter decreases cause of the collision frequency rising, below its critical value which is then about 2. Stable operation with several megawatts in power output had been experimentally achieved as from 1967 with high electron temperature. But this electrothermal control does not allow to decrease T_g low enough for long duration conditions (thermal ablation) so such a solution is not practical for any industrial energy conversion.

Another idea to control the instability would be to increase non-thermal ionisation rate thanks to a laser which would act like a guidance system for streamers between electrodes, increasing the electron density and the conductivity, therefore lowering the Hall parameter under its critical value along these paths. But this concept has never been tested experimentally.

In the 1970s and more recently, some researchers tried to master the instability with oscillating fields. Oscillations of the electric field or of an additional RF electromagnetic field locally modify the Hall parameter.

Finally, a solution has been found in the early 1980s to annihilate completely the electrothermal instability within MHD converters, thanks to non-homogeneous magnetic fields. A strong magnetic field implies a high Hall parameter, therefore a low electrical conductivity in the medium. So the idea is to make some "paths" linking an electrode to the other, where the magnetic field is locally attenuated. Then the electric current tends to flow in these low B-field paths as thin plasma cords or streamers, where the electron density and temperature increase. The plasma becomes locally Coulombian, and the local Hall parameter value falls, while its critical threshold rises. Experiments where streamers do not present any inhomogeneity has been obtained with this method. This effect, strongly nonlinear, was unexpected but led to a very effective system for streamer guidance.

Two-stream Instability

The two-stream instability is a very common instability in plasma physics. It can be induced by an energetic particle stream injected in a plasma, or setting a current along the plasma so different species (ions and electrons) can have different drift velocities. The energy from the particles can lead to plasma wave excitation.

Two-stream instability can arise from the case of two cold beams, in which no particles are resonant with the wave, or from two hot beams, in which there exist particles from one or both beams which are resonant with the wave.

Two-stream instability is known in various limiting cases as beam-plasma instability, beam instability, or bump-on-tail instability.

Dispersion Relation in Cold-beam Limit

Consider a cold, uniform, and unmagnetized plasma, where ions are stationary and the electrons have velocity $\mathbf{V}_0$, that is, the reference frame is moving with the ion stream. Let the electrostatic waves be of the form:

$$\mathbf{E}_1 = \xi_1 \exp[i(kx - \omega t)]\hat{\mathbf{x}}$$

Applying linearization techniques to the equation of motions for both species, to the equation of continuity, and Poisson's equation, and introducing the spatial and temporal harmonic operators $\partial_t \rightarrow -i\omega \nabla \rightarrow ik$ we can get the following expression:

$$1 = \omega_{pe}^2 \left[\frac{m_e / m_i}{\omega^2} + \frac{1}{(\omega - kv_0)^2} \right],$$

which represents the dispersion relation for longitudinal waves, and represents a quartic equation in ω.. The roots can be expressed in the form:

$$\omega_j = \omega_j^R + i\gamma_j$$

If the imaginary part ($Im(\omega_j)$) is zero, then the solutions represent all the possible modes, and there is no temporal wave growth or damping at all:

$$\mathbf{E} = \xi \exp[i(kx - \omega t)]\hat{\mathbf{x}}$$

If $Im(\omega_j) \neq 0$, that is, any of the roots are complex, they will occur in complex conjugate pairs. Substituting in the expression for electrostatic waves leads to:

$$\mathbf{E} = \xi \exp[i(kx - \omega_j^R t)] \exp[\gamma t]\hat{\mathbf{x}}$$

Because of the second exponential function at the right, the temporal dynamics of the wave amplitude depends strongly on the parameter γ; if $\gamma < 0$, then the waves will be exponentially damped; on the other hand, if $\gamma < 0$, then the waves are unstable and will grow at an exponential rate.

Wave–particle Interactions

In the hot-beam case, the two-stream instability can be thought of as the inverse of Landau damping. There are particles which have the same velocity as the wave. The existence of a greater number of particles that move slower than the wave phase velocity $^{V}{}_{ph}$ as compared with those that move faster, leads to an energy transfer from the wave to the particles. In the case of the two-stream instability, when an electron stream is injected to the plasma, the particles' velocity distribution function has a "bump" on its "tail". If a wave has phase velocity in the region where the slope is positive, there is a greater number of faster particles ($v > v_{ph}$) than slower particles, and so there is a greater amount of energy being transferred from the fast particles to the wave, giving rise to exponential wave growth.

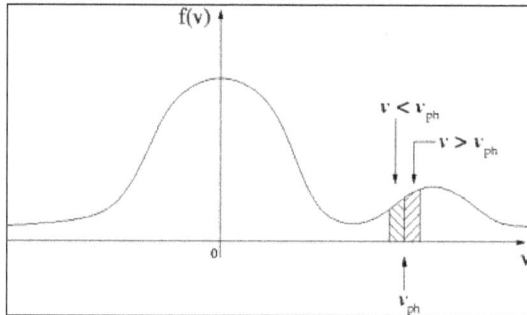

In the cold-beam case, there are no particles which have the same velocity as the phase velocity of the wave (no particles are *resonant*). However, the wave can grow exponentially. In this case, the beam particles are bunched in space in a propagating wave in a self-reinforcing way even though no particles move with the propagation velocity.

In both the hot-beam and cold-beam case, the instability grows until the beam particles are trapped in the electric field of the wave. This is when the instability is said to *saturate*.

Weibel Instability

The Weibel instability is a plasma instability present in homogeneous or nearly homogeneous electromagnetic plasmas which possess an anisotropy in momentum (velocity) space. This anisotropy is most generally understood as two temperatures in different directions. Burton Fried showed that this instability can be understood more simply as the superposition of many counter-streaming beams. In this sense, it is like the two-stream instability except that the perturbations are electromagnetic and result in filamentation as opposed to electrostatic perturbations which would result in charge bunching. In the linear limit the instability causes exponential growth of electromagnetic fields in the plasma which help restore momentum space isotropy. In very extreme cases, the Weibel instability is related to one- or two-dimensional stream instabilities.

Consider an electron-ion plasma in which the ions are fixed and the electrons are hotter in the y-direction than in x or z-direction.

To see how magnetic field perturbation would grow, suppose a field B = B cos kx spontaneously arises from noise. The Lorentz force then bends the electron trajectories with the result that upward-moving-ev x B electrons congregate at B and downward-moving ones at A. The resulting current j = -en ve sheets generate magnetic field that enhances the original field and thus perturbation grows.

Weibel instability is also common in astrophysical plasmas, such as collisionless shock formation in supernova remnants and γ-ray bursts.

A Simple Example of Weibel Instability

As a simple example of Weibel instability, consider an electron beam with density n_{b0} and initial velocity $v_0 \mathbf{z}$ propagating in a plasma of density $v_0 \mathbf{z}$ with velocity $n_{p0} = n_{b0}$. The analysis below will show how an electromagnetic perturbation in the form of a plane wave gives rise to a Weibel instability in this simple anisotropic plasma system. We assume a non-relativistic plasma for simplicity.

We assume there is no background electric or magnetic field i.e. $\mathbf{B}_0 = \mathbf{E}_0 = 0$. The perturbation will be taken as an electromagnetic wave propagating along $\hat{\mathbf{x}}$. i.e. $\mathbf{k} = k\hat{\mathbf{x}}$. Assume the electric field has the form,

$$\mathbf{E}_1 = A e^{i(kx - \omega t)} \mathbf{z}$$

With the assumed spatial and time dependence, we may use $\dfrac{\partial}{\partial t} \to -i\omega$ and $\nabla \to ik\hat{\mathbf{x}}$.

From Faraday's Law, we may obtain the perturbation magnetic field,

$$\nabla \times \mathbf{E}_1 = -\frac{\partial \mathbf{B}_1}{\partial t} \Rightarrow i\mathbf{k} \times \mathbf{E}_1 = i\omega \mathbf{B}_1 \Rightarrow \mathbf{B}_1 = \hat{\mathbf{y}} \frac{k}{\omega} E_1$$

Consider the electron beam. We assume small perturbations, and so linearize the velocity $\mathbf{v}_b = \mathbf{v}_{b0} + \mathbf{v}_{b1}$ and density $n_b = n_{b0} + n_{b1}$. The goal is to find the perturbation electron beam current density,

$$\mathbf{J}_{b1} = -en_b \mathbf{v}_b = -en_{b0} \mathbf{v}_{b1} - en_{b1} \mathbf{v}_{b0}$$

where second-order terms have been neglected. To do that, we start with the fluid momentum equation for the electron beam,

$$m(\frac{\partial \mathbf{v}_b}{\partial t} + (\mathbf{v}_b \cdot \nabla)\mathbf{v}_b) = -e\mathbf{E} - e\mathbf{v}_b \times \mathbf{B}$$

which can be simplified by noting that $\dfrac{\partial \mathbf{v}_{b0}}{\partial t} = \nabla \cdot v_{b0} = 0$ and neglecting second-order terms. With the plane wave assumption for the derivatives, the momentum equation becomes,

$$-i\omega m \mathbf{v}_{b1} = -e\mathbf{E}_1 - e\mathbf{v}_{b0} \times \mathbf{B}_1$$

We can decompose the above equations in components, paying attention to the cross product at the far right, and obtain the non-zero components of the beam velocity perturbation:

$$v_{b1z} = \frac{eE_1}{mi\omega}$$

$$v_{b1x} = \frac{eE_1}{mi\omega} \frac{kv_{b0}}{\omega}$$

To find the perturbation density n_{b1}, we use the fluid continuity equation for the electron beam,

$$\frac{n_b}{\partial t} + \nabla \cdot (n_b \mathbf{v_b}) = 0$$

which can again be simplified by noting that $\dfrac{\partial n_{b0}}{\partial t} = \nabla n_{b0} = 0$ and neglecting second-order terms. The result is,

$$n_{b1} = n_{b0}\frac{k}{\omega} v_{b1x}$$

Using these results, we may use the equation for the beam perturbation current density given above to find,

$$J_{b1x} = -n_{b0}e^2 E_1 \frac{kv_{b0}}{im\omega^2}$$

$$J_{b1z} = -n_{b0}e^2 E_1 \frac{1}{im\omega}(1 + \frac{k^2 v_{b0}^2}{\omega^2})$$

Analogous expressions can be written for the perturbation current density of the left-moving plasma. By noting that the x-component of the perturbation current density is proportional to v_o, we see that with our assumptions for the beam and plasma unperturbed densities and velocities the x-component of the net current density will vanish, whereas the z-components, which are proportional to v_o^2, will add. The net current density perturbation is therefore,

$$\mathbf{J}_1 = -2n_{b0}e^2 E_1 \frac{1}{im\omega}(1 + \frac{k^2 v_{b0}^2}{\omega^2})\hat{\mathbf{z}}$$

The dispersion relation can now be found from Maxwell's Equations:

$$\nabla \times \mathbf{E}_1 = i\omega \mathbf{B}_1$$

$$\nabla \times \mathbf{B}_1 = \mu_0 \mathbf{J}_1 - i\omega\epsilon_0\mu_0 \mathbf{E}_1$$

$$\Rightarrow \nabla \times \nabla \times \mathbf{E}_1 = -\nabla^2 \mathbf{E}_1 + \nabla(\nabla \cdot \mathbf{E}_1) = k^2 \mathbf{E}_1 + i\mathbf{k}(i\mathbf{k} \cdot \mathbf{E}_1) = k^2 \mathbf{E}_1 = i\omega\nabla \times \mathbf{B}_1$$

$$= \frac{i\omega}{c^2\epsilon_0}\mathbf{J}_1 + \frac{\omega^2}{c^2}\mathbf{E}_1$$

where $c = \dfrac{1}{\sqrt{\epsilon_0\mu_0}}$ is the speed of light in free space. By defining the effective plasma, frequency $\omega_p^2 = \dfrac{2n_{b0}e^2}{\epsilon_0 m}$,, the equation above results in,

$$k^2 - \frac{\omega^2}{c^2} = -\frac{\omega_p^2}{c^2}(1+\frac{k^2 v_0^2}{\omega^2}) \Rightarrow \omega^4 - \omega^2(\omega_p^2 + k^2 c^2) - \omega_p^2 k^2 v_0^2 = 0$$

This bi-quadratic equation may be easily solved to give the dispersion relation,

$$\omega^2 = \frac{1}{2}(\omega_p^2 + k^2 c^2 \pm \sqrt{(\omega_p^2 + k^2 c^2)^2 + 4\omega_p^2 k^2 v_0^2})$$

In the search for instabilities, we look for $k\,Im(\omega) \neq 0$ (is assumed real). Therefore, we must take the dispersion relation/mode corresponding to the minus sign in the equation above.

To gain further insight on the instability, it is useful to harness our non-relativistic assumption >> to simplify the square root term, by noting that,

$$\sqrt{(\omega_p^2 + k^2 c^2)^2 + 4\omega_p^2 k^2 v_0^2} = (\omega_p^2 + k^2 c^2)(1 + \frac{4\omega_p^2 k^2 v_0^2}{(\omega_p^2 + k^2 c^2)^2})^{1/2}$$

$$\approx (\omega_p^2 + k^2 c^2)(1 + \frac{2\omega_p^2 k^2 v_0^2}{(\omega_p^2 + k^2 c^2)^2})$$

The resulting dispersion relation is then much simpler,

$$\omega^2 = \frac{-\omega_p^2 k^2 v_0^2}{\omega_p^2 + k^2 c^2} < 0$$

ω is purely imaginary. Writing $\omega = i\gamma$

$$\gamma = \frac{\omega_p k v_0}{(\omega_p^2 + k^2 c^2)^{1/2}} = \omega_p \frac{v_0}{c} \frac{1}{(1 + \frac{\omega_p^2}{k^2 c^2})^{1/2}}$$

we see that $Im(\omega) > 0$, indeed corresponding to an instability.

The electromagnetic fields then have the form,

$$\mathbf{E}_1 = A\hat{z}e^{\gamma t}e^{ikx}$$

$$\mathbf{B}_1 = \hat{y}\frac{k}{\omega}E_1 = \hat{y}\frac{k}{i\gamma}Ae^{\gamma t}e^{ikx}$$

Therefore, the electric and magnetic fields are 90^0 out of phase, and by noting that,

$$\frac{|B_1|}{|E_1|} = \frac{k}{\gamma} \propto \frac{c}{v_0} \gg 1$$

so we see this is a primarily magnetic perturbation although there is a non-zero electric perturbation. The magnetic field growth results in the characteristic filamentation structure of Weibel instability. Saturation will happen when the growth rate γ is on the order of the electron cyclotron frequency,

$$\gamma \sim \omega_p \frac{v_0}{c} \sim \omega_c \Rightarrow B \sim \frac{m}{e} \omega_p \frac{v_0}{c}$$

CORONA DISCHARGE

A corona discharge is an electrical discharge brought on by the ionization of a fluid such as air surrounding a conductor that is electrically charged. Spontaneous corona discharges occur naturally in high-voltage systems unless care is taken to limit the electric field strength. A corona will occur when the strength of the electric field (potential gradient) around a conductor is high enough to form a conductive region, but not high enough to cause electrical breakdown or arcing to nearby objects. It is often seen as a bluish (or another color) glow in the air adjacent to pointed metal conductors carrying high voltages, and emits light by the same property as a gas discharge lamp.

In many high voltage applications, corona is an unwanted side effect. Corona discharge from high voltage electric power transmission lines constitutes an economically significant waste of energy for utilities. In high voltage equipment like Cathode Ray Tube televisions, radio transmitters, X-ray machines, and particle accelerators the current leakage caused by coronas can constitute an unwanted load on the circuit. In the air, coronas generate gases such as ozone (O_3) and nitric oxide (NO), and in turn, nitrogen dioxide (NO_2), and thus nitric acid (HNO_3) if water vapor is present. These gases are corrosive and can degrade and embrittle nearby materials, and are also toxic to humans and the environment.

Corona discharges can often be suppressed by improved insulation, corona rings, and making high voltage electrodes in smooth rounded shapes. However, controlled corona discharges are used in a variety of processes such as air filtration, photocopiers, and ozone generators.

A corona discharge is a process by which a current flows from an electrode with a high potential into a neutral fluid, usually air, by ionizing that fluid so as to create a region of plasma around the electrode. The ions generated eventually pass the charge to nearby areas of lower potential, or recombine to form neutral gas molecules.

When the potential gradient (electric field) is large enough at a point in the fluid, the fluid at that point ionizes and it becomes conductive. If a charged object has a sharp point, the electric field strength around that point will be much higher than elsewhere. Air near the electrode can become ionized (partially conductive), while regions more distant do not. When the air near the point becomes conductive, it has the effect of increasing the apparent size of the conductor. Since the new conductive region is less sharp, the ionization may not extend past this local region. Outside this region of ionization and conductivity, the charged particles slowly find their way to an oppositely charged object and are neutralized.

Along with the similar brush discharge, the corona is often called a "single-electrode discharge", as opposed to a "two-electrode discharge"; an electric arc. A corona only forms when the conductor is widely enough separated from conductors at the opposite potential that an arc cannot jump between them. If the geometry and gradient are such that the ionized region continues to grow until it reaches another conductor at a lower potential, a low resistance conductive path between the two will be formed, resulting in an electric spark or electric arc, depending upon the source of the electric field. If the source continues to supply current, a spark will evolve into a continuous discharge called an arc.

Corona discharge only forms when the electric field (potential gradient) at the surface of the conductor exceeds a critical value, the dielectric strength or disruptive potential gradient of the fluid. In air at atmospheric pressure, it is roughly 30 kilovolts per centimeter but decreases with pressure, so Corona is more of a problem at high altitudes. Corona discharge usually forms at highly curved regions on electrodes, such as sharp corners, projecting points, edges of metal surfaces, or small diameter wires. The high curvature causes a high potential gradient at these locations so that the air breaks down and forms plasma there first. On sharp points in air corona can start at potentials of 2 - 6 kV. In order to suppress corona formation, terminals on high voltage equipment are frequently designed with smooth large diameters rounded shapes like balls or toruses, and corona rings are often added to insulators of high voltage transmission lines.

Coronas may be *positive* or *negative*. This is determined by the polarity of the voltage on the highly curved electrode. If the curved electrode is positive with respect to the flat electrode, it has a *positive corona*, if it is negative, it has a *negative corona*. The physics of positive and negative coronas are strikingly different. This asymmetry is a result of the great difference in mass between electrons and positively charged ions, with only the electron having the ability to undergo a significant degree of ionizing inelastic collision at common temperatures and pressures.

An important reason for considering coronas is the production of ozone around conductors undergoing corona processes in air. A negative corona generates much more ozone than the corresponding positive corona.

Large corona discharges *(white)* around conductors energized
by a 1.05 million volt transformer.

Applications

Corona discharge has a number of commercial and industrial applications:

- Removal of unwanted electric charges from the surface of aircraft in flight and thus avoiding the detrimental effect of uncontrolled electrical discharge pulses on the performance of avionic systems.

- Manufacture of ozone.

- Sanitization of pool water.

- In an electrostatic precipitator, removal of solid pollutants from a waste gas stream, or scrubbing particles from the air in air-conditioning systems.

- Photocopying.

- Air ionisers.

- Production of photons for Kirlian photography to expose photographic film.

- EHD thrusters, Lifters, and other ionic wind devices.

- Nitrogen laser.

- Ionization of a gaseous sample for subsequent analysis in a mass spectrometer or an ion mobility spectrometer.

- Static charge neutralization, as applied through antistatic devices like ionizing bars.

Coronas can be used to generate charged surfaces, which is an effect used in electrostatic copying (photocopying). They can also be used to remove particulate matter from air streams by first charging the air, and then passing the charged stream through a comb of alternating polarity, to deposit the charged particles onto oppositely charged plates.

The free radicals and ions generated in corona reactions can be used to scrub the air of certain noxious products, through chemical reactions, and can be used to produce ozone.

Problems

Coronas can generate audible and radio-frequency noise, particularly near electric power transmission lines. Therefore, power transmission equipment is designed to minimize the formation of corona discharge.

Corona discharge is generally undesirable in:

- Electric power transmission, where it causes:
  - Power loss.
  - Audible noise.
  - Electromagnetic interference.
  - Purple Glow.
  - Ozone production.
  - Insulation damage.
  - Possible distress in animals that are sensitive to ultraviolet light.
- Electrical components such as transformers, capacitors, electric motors, and generators:
  - Corona can progressively damage the insulation inside these devices, leading to equipment failure.
  - Elastomer items such as O-rings can suffer ozone cracking.
  - Plastic film capacitors operating at mains voltage can suffer progressive loss of capacitance as corona discharges cause local vaporization of the metallization.

In many cases, coronas can be suppressed by corona rings, toroidal devices that serve to spread the electric field over a larger areas and decrease the field gradient below the corona threshold.

Mechanism

Corona discharge results when the electric field is strong enough to create a chain reaction: electrons in the air collide with atoms hard enough to ionize them, creating more free electrons which ionize more atoms. The diagrams below illustrate at a microscopic scale the process which creates a corona in the air next to a pointed electrode carrying a high negative voltage with respect to ground. The process is:

- A neutral atom or molecule, in a region of the strong electric field (such as the high potential gradient near the curved electrode), is ionized by a natural environmental event (for example, being struck by an ultraviolet photon or cosmic ray particle), to create a positive ion and a free electron.

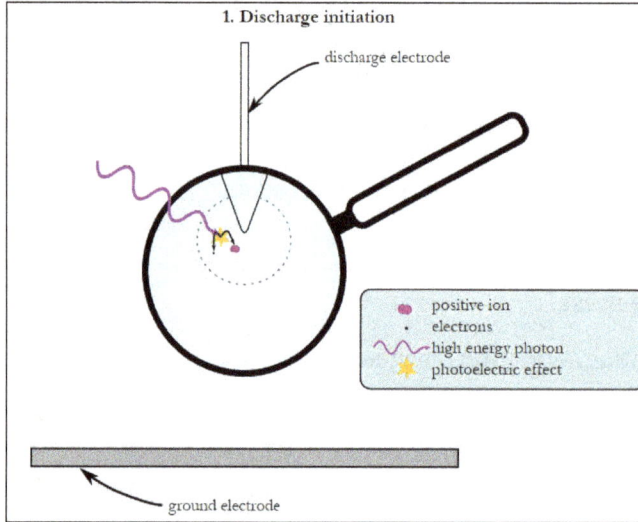

- The electric field accelerates these oppositely charged particles in opposite directions, separating them, preventing their recombination, and imparting kinetic energy to each of them.

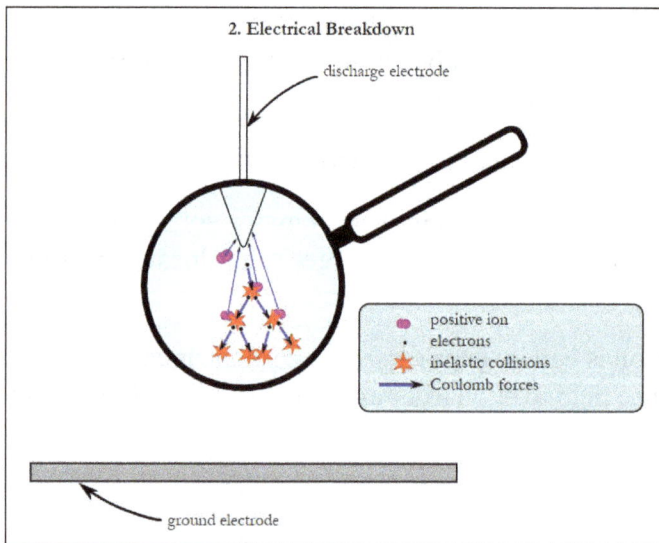

- The electron has a much higher charge/mass ratio and so is accelerated to a higher velocity than the positive ion. It gains enough energy from the field that when it strikes another atom it ionizes it, knocking out another electron, and creating another positive ion. These electrons are accelerated and collide with other atoms, creating further electron/positive-ion pairs, and these electrons collide with more atoms, in a chain reaction process called an electron avalanche. Both positive and negative coronas rely on electron avalanches. In a positive corona, all the electrons are attracted inward toward the nearby positive electrode and the ions are repelled outwards. In a negative corona, the ions are attracted inward and the electrons are repelled outwards.

- The glow of the corona is caused by electrons recombining with positive ions to form neutral atoms. When the electron falls back to its original energy level, it releases a photon of light. The photons serve to ionize other atoms, maintaining the creation of electron avalanches.

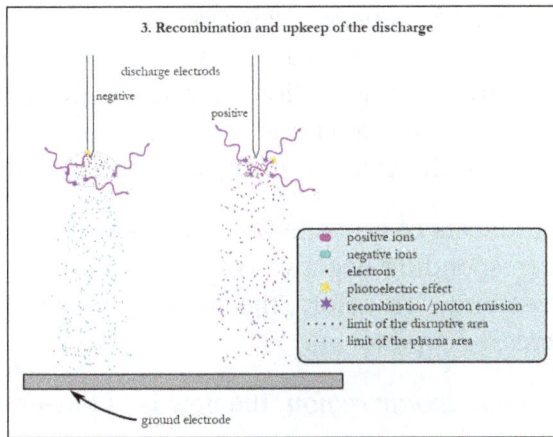

3. Recombination and upkeep of the discharge

- At a certain distance from the electrode, the electric field becomes low enough that it no longer imparts enough energy to the electrons to ionize atoms when they collide. This is the outer edge of the corona. Outside this, the ions move through the air without creating new ions. The outward moving ions are attracted to the opposite electrode and eventually reach it and combine with electrons from the electrode to become neutral atoms again, completing the circuit.

Thermodynamically, a corona is a very nonequilibrium process, creating a non-thermal plasma. The avalanche mechanism does not release enough energy to heat the gas in the corona region generally and ionize it, as occurs in an electric arc or spark. Only a small number of gas molecules take part in the electron avalanches and are ionized, having energies close to the ionization energy of 1–3 ev, the rest of the surrounding gas is close to ambient temperature.

The onset voltage of corona or corona inception voltage (CIV) can be found with *Peek's law*, formulated from empirical observations. Later papers derived more accurate formulas.

Positive Coronas

Properties

A positive corona is manifested as a uniform plasma across the length of a conductor. It can often be seen glowing blue/white, though many of the emissions are in the ultra-violet. The uniformity of the plasma is caused by the homogeneous source of secondary avalanche electrons.With the same geometry and voltages, it appears a little smaller than the corresponding negative corona, owing to the lack of a non-ionising plasma region between the inner and outer regions.

A positive corona has a much lower density of free electrons compared to a negative corona; perhaps a thousandth of the electron density, and a hundredth of the total number of electrons. However, the electrons in a positive corona are concentrated close to the surface of the curved conductor, in a region of the high potential gradient (and therefore the electrons have high energy), whereas in a negative corona many of the electrons are in the outer, lower-field areas. Therefore, if electrons are to be used in an application which requires high activation energy, positive coronas may support a greater reaction constant than corresponding negative coronas; though the total number of electrons may be lower, the number of very high energy electrons may be higher.

Coronas are efficient producers of ozone in the air. A positive corona generates much less ozone than the corresponding negative corona, as the reactions which produce ozone are relatively low-energy. Therefore, the greater number of electrons of a negative corona leads to increased production.

Beyond the plasma, in the *unipolar region*, the flow is of low-energy positive ions toward the flat electrode.

Mechanism

As with a negative corona, a positive corona is initiated by an exogenous ionization event in a region of a high potential gradient. The electrons resulting from the ionization are attracted *toward* the curved electrode, and the positive ions repelled from it. By undergoing inelastic collisions closer and closer to the curved electrode, further molecules are ionized in an electron avalanche.

In a positive corona, secondary electrons, for further avalanches, are generated predominantly in the fluid itself, in the region outside the plasma or avalanche region. They are created by ionization caused by the photons emitted from that plasma in the various de-excitation processes occurring within the plasma after electron collisions, the thermal energy liberated in those collisions creating photons which are radiated into the gas. The electrons resulting from the ionization of a neutral gas molecule are then electrically attracted back toward the curved electrode, attracted *into* the plasma, and so begins the process of creating further avalanches inside the plasma.

Negative Coronas

Properties

A negative corona is manifested in a non-uniform corona, varying according to the surface features and irregularities of the curved conductor. It often appears as tufts of the corona at sharp edges, the number of tufts altering with the strength of the field. The form of negative coronas is a result of its source of secondary avalanche electrons. It appears a little larger than the corresponding positive corona, as electrons are allowed to drift out of the ionizing region, and so the plasma continues some distance beyond it. The total number of electrons and electron density is much greater than in the corresponding positive corona. However, they are of predominantly lower energy, owing to being in a region of lower potential gradient. Therefore, whilst for many reactions, the increased electron density will increase the reaction rate, the lower energy of the electrons will mean that reactions which require higher electron energy may take place at a lower rate.

Mechanism

Negative coronas are more complex than positive coronas in construction. As with positive coronas, the establishing of a corona begins with an exogenous ionization event generating a primary electron, followed by an electron avalanche.

Electrons ionized from the neutral gas are not useful in sustaining the negative corona process by generating secondary electrons for further avalanches, as the general movement of electrons in a negative corona is outward from the curved electrode. For negative corona, instead, the dominant process generating secondary electrons is the photoelectric effect, from the surface of the electrode itself. The work function of the electrons (the energy required to liberate the electrons from the surface) is considerably lower than the ionization energy of air at standard temperatures and pressures, making it a more liberal source of secondary electrons under these conditions. Again, the source of energy for the electron-liberation is a high-energy photon from an atom within the plasma body relaxing after excitation from an earlier collision. The use of ionized neutral gas as a source of ionization is further diminished in a negative corona by the high-concentration of positive ions clustering around the curved electrode.

Under other conditions, the collision of the positive species with the curved electrode can also cause electron liberation.

The difference, then, between positive and negative coronas, in the matter of the generation of secondary electron avalanches, is that in a positive corona they are generated by the gas surrounding the plasma region, the new secondary electrons travelling inward, whereas in a negative corona they are generated by the curved electrode itself, the new secondary electrons travelling outward.

A further feature of the structure of negative coronas is that as the electrons drift outwards, they encounter neutral molecules and, with electronegative molecules (such as oxygen and water vapor), combine to produce negative ions. These negative ions are then attracted to the positive uncurved electrode, completing the 'circuit'.

Electrical Wind

Ionized gases produced in a corona discharge are accelerated by the electric field, producing a movement of gas or *electrical wind*. The air movement associated with a discharge current of a few hundred microamperes can blow out a small candle flame within about 1 cm of a discharge point. A pinwheel, with radial metal spokes and pointed tips bent to point along the circumference of a circle, can be made to rotate if energized by a corona discharge; the rotation is due to the differential electric attraction between the metal spokes and the space charge shield region that surrounds the tips.

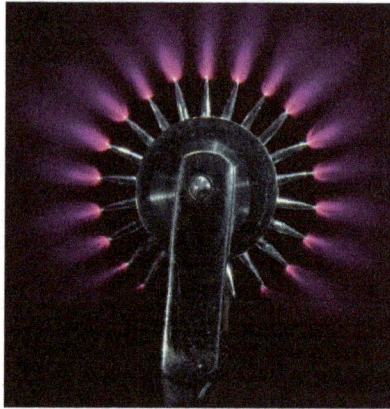

Corona discharge on a Wartenberg wheel.

CORONAL SEISMOLOGY

Coronal seismology is a technique of studying the plasma of the Sun's corona with the use of magnetohydrodynamic (MHD) waves and oscillations. Magnetohydrodynamics studies the dynamics of electrically conducting fluids - in this case the fluid is the coronal plasma. Observed properties of the waves (e.g. period, wavelength, amplitude, temporal and spatial signatures (what is the shape of the wave perturbation?), characteristic scenarios of the wave evolution (is the wave damped?), combined with a theoretical modelling of the wave phenomena (dispersion relations, evolutionary equations, etc.), may reflect physical parameters of the corona which are not accessible *in situ,* such as the coronal magnetic field strength and Alfvén velocity and coronal dissipative coefficients. Originally, the method of MHD coronal seismology was suggested by Y. Uchida in 1970 for propagating waves, and B. Roberts et al. in 1984 for standing waves, but was not practically applied until the late 90s due to a lack of necessary observational resolution. Philosophically, coronal seismology is similar to the Earth's seismology, helioseismology, and MHD spectroscopy of laboratory plasma devices. In all these approaches, waves of various kind are used to probe a medium.

The theoretical foundation of coronal seismology is the dispersion relation of MHD modes of a plasma cylinder: a plasma structure which is nonuniform in the transverse direction and extended along the magnetic field. This model works well for the description of a number of plasma structures observed in the solar corona: e.g. coronal loops, prominence fibrils, plumes, various filaments. Such a structure acts as a waveguide of MHD waves.

Types of Magnetohydrodynamic Waves

There are several distinct kinds of MHD modes which have quite different dispersive, polarisation, and propagation properties:

- Kink (or transverse) modes, which are oblique fast magnetoacoustic (also known as magnetosonic waves) guided by the plasma structure; the mode causes the displacement of the axis of the plasma structure. These modes are weakly compressible, but could nevertheless be observed with imaging instruments as periodic standing or propagating displacements of coronal structures, e.g. coronal loops. The frequency of transverse or "kink" modes is given by the following expression:

$$\omega_K = \sqrt{\frac{2k_z B^2}{\mu(\rho_i + \rho_e)}}$$

For kink modes the parameter the azimuthal wave number in a cylindrical model of a loop, is equal to 1, meaning that the cylinder is swaying with fixed ends.

- Sausage modes, which are also oblique fast magnetoacoustic waves guided by the plasma structure; the mode causes expansions and contractions of the plasma structure, but does not displace its axis. These modes are compressible and cause significant variation of the absolute value of the magnetic field in the oscillating structure. The frequency of sausage modes is given by the following expression:

$$\omega_S = \sqrt{\frac{k_z^2 B^2}{\mu \rho_e}}$$

For sausage modes the parameter m is equal to (0); this would be interpreted as a "breathing" in and out, again with fixed endpoints.

- Longitudinal (or slow, or acoustic) modes, which are slow magnetoacoustic waves propagating mainly along the magnetic field in the plasma structure;

these mode are essentially compressible. The magnetic field perturbation in these modes is negligible. The frequency of slow modes is given by the following expression:

$$\omega_L = \sqrt{k_z^2 \left(\frac{C_s^2 C_A^2}{C_s^2 + C_A^2} \right)}$$

Where we define C_s as the sound speed and C_A as the Alfvén velocity.

- Torsional (Alfvén or twist) modes are incompressible transverse perturbations of the magnetic field along certain individual magnetic surfaces. In contrast with kink modes, torsional modes cannot be observed with imaging instruments, as they do not cause the displacement of either the structure axis or its boundary.

$$\omega_A = \sqrt{\frac{k_z^2 B^2}{\mu \rho_i}}$$

Observations

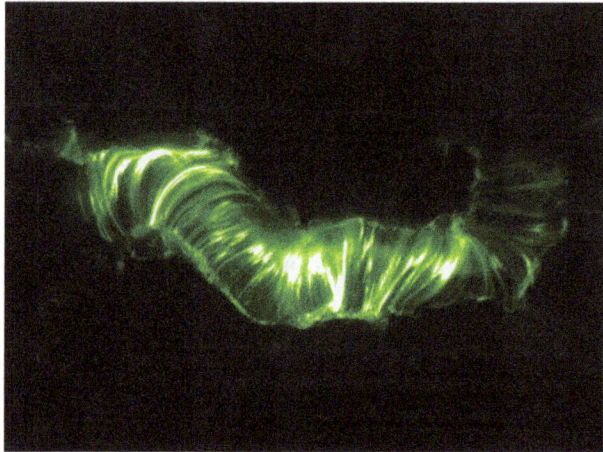

TRACE image of a coronal arcade.

Wave and oscillatory phenomena are observed in the hot plasma of the corona mainly in EUV, optical and microwave bands with a number of spaceborne and ground-based instruments, e.g. the Solar and Heliospheric Observatory (SOHO), the Transition Region and Coronal Explorer (TRACE), the Nobeyama Radioheliograph (NoRH, see the Nobeyama radio observatory). Phenomenologically, researchers distinguish between compressible waves in polar plumes and in legs of large coronal loops, flare-generated transverse oscillations of loops, acoustic oscillations of loops, propagating kink waves in loops and in structures above arcades (an arcade being a close collection of loops in a cylindrical structure), sausage oscillations of flaring loops, and oscillations of prominences and fibrils, and this list is continuously updated.

Coronal seismology is one of the aims of the Atmospheric Imaging Assembly (AIA) instrument on the Solar Dynamics Observatory (SDO) mission.

The potential of coronal seismology in the estimation of the coronal magnetic field, density scale height, "fine structure" (by which is meant the variation in structure of an inhomogeneous structure such as an inhomogeneous coronal loop) and heating has been demonstrated by different research groups. Work relating to the coronal magnetic field was mentioned earlier. It has been shown that sufficiently broadband slow magnetoacoustic waves, consistent with currently available observations in the low frequency part of the spectrum, could provide the rate of heat deposition sufficient to heat a coronal loop. Regarding the density scale height, transverse oscillations of coronal loops that have both variable circular cross-sectional area and plasma density in the longitudinal direction have been studied theoretically. A second order ordinary differential equation has been derived describing the displacement of the loop axis. Together with boundary conditions, solving this equation determines the eigenfrequencies and eigenmodes. The coronal density scale height could then be estimated by using the observed ratio of the fundamental frequency and first overtone of loop kink oscillations. Little is known of coronal fine structure. Doppler shift oscillations in hot active region loops obtained with the Solar Ultraviolet Measurements of Emitted Radiation instrument (SUMER) aboard SOHO have been studied. The spectra were recorded along a 300 arcsec slit placed at a fixed position in the corona above the active regions. Some oscillations showed phase propagation along the slit in one or both directions with apparent speeds in the range of 8–102 km per second, together with distinctly different intensity and line width distributions along the slit. These features can be explained by the excitation of the oscillation at a footpoint of an inhomogeneous coronal loop, e.g. a loop with fine structure.

DIFFUSION DAMPING

In modern cosmological theory, diffusion damping, also called photon diffusion damping, is a physical process which reduced density inequalities (anisotropies) in the early universe, making the universe itself and the cosmic microwave background radiation (CMB) more uniform. Around 300,000 years after the Big Bang, during the epoch of recombination, diffusing photons travelled from hot regions of space to cold ones, equalising the temperatures of these regions. This effect is responsible, along with baryon acoustic oscillations, the Doppler effect, and the effects of gravity on electromagnetic radiation, for the eventual formation of galaxies and galaxy clusters, these being the dominant large scale structures which are observed in the universe. It is a damping by diffusion, not of diffusion.

The strength of diffusion damping is calculated by a mathematical expression for the damping factor, which figures into the Boltzmann equation, an equation which describes the amplitude of perturbations in the CMB. The strength of the diffusion damping is chiefly governed by the distance photons travel before being scattered (diffusion length). The primary effects on the diffusion length are from the properties of the plasma in question: different sorts of plasma may experience different sorts of diffusion damping. The evolution of a plasma may also affect the damping process. The scale on which diffusion damping works is called the Silk scale and its value corresponds to the size of galaxies of the present day. The mass contained within the Silk scale is called the Silk mass and it corresponds to the mass of the galaxies.

Diffusion damping took place about 13.8 billion years ago, during the stage of the early universe called recombination or matter-radiation decoupling. This period occurred about 320,000 years after the Big Bang. This is equivalent to a redshift of around $z = 1090$. Recombination was the stage during which simple atoms, e.g. hydrogen and helium, began to form in the cooling, but still very hot, soup of protons, electrons and photons that composed the universe. Prior to the recombination epoch, this soup, a plasma, was largely opaque to the electromagnetic radiation of photons. This meant that the permanently excited photons were scattered by the protons and electrons too often to travel very far in straight lines. During the recombination epoch, the universe cooled rapidly as free electrons were captured by atomic nuclei; atoms formed from their constituent parts and the universe became transparent: the amount of photon scattering decreased dramatically. Scattering less, photons could diffuse (travel) much greater distances. There was no significant diffusion damping for electrons, which could not diffuse nearly as far as photons could in similar circumstances. Thus all damping by electron diffusion is negligible when compared to photon diffusion damping.

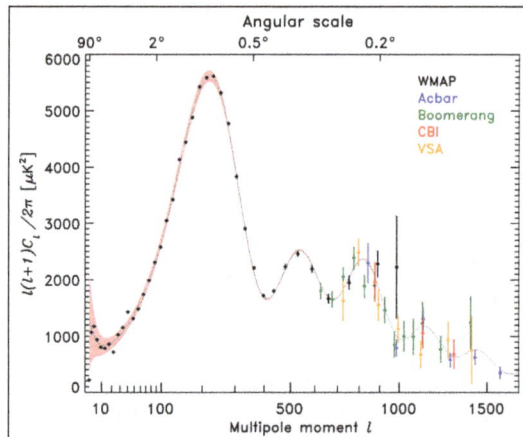

The power spectrum of the cosmic microwave background radiation
temperature anisotropy in terms of the angular scale (or multipole moment).
Diffusion damping can be easily seen in the suppression of power peaks when $l \gtrsim 1000$.

Acoustic perturbations of initial density fluctuations in the universe made some regions of space hotter and denser than others. These differences in temperature and density are called anisotropies. Photons diffused from the hot, overdense regions of plasma to the cold, underdense ones: they dragged along the protons and electrons: the photons pushed electrons along, and these, in turn, pulled on protons by the Coulomb force. This caused the temperatures and densities of the hot and cold regions to be averaged and the universe became less anisotropic (characteristically various) and more isotropic (characteristically uniform). This reduction in anisotropy is the damping of diffusion damping. Diffusion damping thus damps temperature and density anisotropies in the early universe. With baryonic matter (protons and electrons) escaping the dense areas along with the photons; the temperature and density inequalities were adiabatically damped. That is to say the ratios of photons to baryons remained constant during the damping process.

Photon diffusion was first described in Joseph Silk's 1968 paper entitled "Cosmic Black-Body Radiation and Galaxy Formation", which was published in The Astrophysical Journal. As such, diffusion damping is sometimes also called Silk damping, though this term may apply only to one possible damping scenario. Silk damping was thus named after its discoverer.

Magnitude

The magnitude of diffusion damping is calculated as a damping factor or suppression factor, represented by the symbol D, which figures into the Boltzmann equation, an equation which describes the amplitude of perturbations in the CMB. The strength of the diffusion damping is chiefly governed by the distance photons travel before being scattered (diffusion length). What affect the diffusion length are primarily the properties of the plasma in question: different sorts of plasma may experience different sorts of diffusion damping. The evolution of a plasma may also affect the damping process.

$$\mathcal{D}(k) = \int_0^{\eta_0} \dot{\tau} e^{-[k/k_D(\eta)]^2} \, d\eta.$$

Where:

- η is the conformal time.

- $\dot{\tau}$ is the "differential optical depth for Thomson scattering". Thomson scattering is the scattering of electromagnetic radiation (light) by charged particles such as electrons.

- k is the wave number of the wave being suppressed.

- $(\dot{\tau} e^{-[k/k_D(\eta)]^2})$ is the visibility function.

  $$k_D(\eta) = 2\pi / \lambda_D$$

The damping factor D, when factored into the Boltzmann equation for the cosmic microwave background radiation (CMB), reduces the amplitude of perturbations:

$$[\Theta_0 + \Psi](\eta_*) = [\hat{\Theta}_0 + \Psi](\eta_*)\mathcal{D}(k).$$

where:

- η_* is the conformal time at decoupling.

- Θ_0 is the "monopole [perturbation] of the photon distribution function"

- Ψ is a "gravitational-potential [perturbation] in the Newtonian gauge". The Newtonian gauge is a quantity with importance in the General Theory of Relativity.

- $[\Theta_0 + \Psi](\eta)$ is the effective temperature.

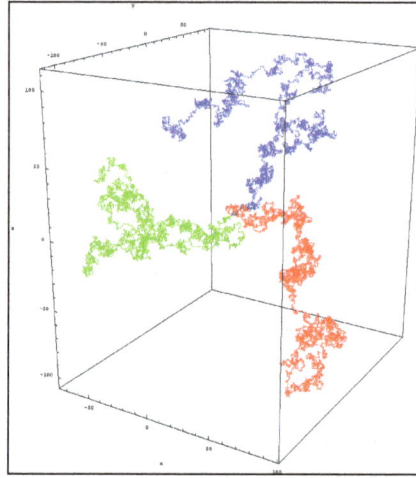

Three random walks in three dimensions.

In diffusion damping, photons from hot regions diffuse to cold regions by random walk, so after steps, the photons have travelled a distance.

Mathematical calculations of the damping factor depend on k_D, or the effective diffusion scale, which in turn depends on a crucial value, the diffusion length, λ_D. The diffusion length relates how far photons travel during diffusion, and comprises a finite number of short steps in random directions. The average of these steps is the Compton mean free path, and is denoted by λ_C. As the direction of these steps are randomly taken, λ_D is approximately equal to $\sqrt{N}\lambda_C$, where N is the number of steps the photon takes before the conformal time at decoupling (η_*).

The diffusion length increases at recombination because the mean free path does, with less photon scattering occurring; this increases the amount of diffusion and damping. The mean free path increases because the electron ionisation fraction, x_e, decreases as ionised hydrogen and helium bind with the free, charged electrons. As this occurs, the mean free path increases proportionally: $\lambda_C \propto (x_e n_b)^{-1}$. That is, the mean free path of the photons is inversely proportional to the electron ionisation fraction and the baryon number density (n_b). That means that the more baryons there were, and the more they were ionised, the shorter the average photon could travel before encountering one and being scattered. Small changes to these values before or during recombination can augment the damping effect considerably. This dependence on the baryon density by photon diffusion allows scientists to use analysis of the latter to investigate the former, in addition to the history of ionisation.

The effect of diffusion damping is greatly augmented by the finite width of the surface of last scattering (SLS). The finite width of the SLS means the CMB photons we see were not all emitted at the same time, and the fluctuations we see are not all in phase. It also means that during recombination, the diffusion length changed dramatically, as the ionisation fraction shifted.

Model Dependence

In general, diffusion damping produces its effects independent of the cosmological model being studied, thereby masking the effects of other, model-*dependent* phenomena. This means that without an accurate model of diffusion damping, scientists cannot judge the relative merits of cosmological models, whose theoretical predictions cannot be compared with observational data, this data being obscured by damping effects. For example, the peaks in the power spectrum due to acoustic oscillations are decreased in amplitude by diffusion damping. This deamplification of the power spectrum hides features of the curve, features that would otherwise be more visible.

Though general diffusion damping can damp perturbations in collisionless dark matter simply due to photon dispersion, the term *Silk damping* applies only to damping of adiabatic models of baryonic matter, which is coupled to the diffusing photons, not dark matter, and diffuses with them. Silk damping is not as significant in models of cosmological development which posit early isocurvature fluctuations (i.e. fluctuations which do not require a constant ratio of baryons and photons). In this case, increases in baryon density do not require a corresponding increases in photon density, and the lower the photon density, the less diffusion there would be: the less diffusion, the less damping. Photon diffusion is not dependent on the causes of the initial fluctuations in the density of the universe.

Effects

Speed

Damping occurs at two different scales, with the process working more quickly over short ranges than over longer distances. Here, a short length is one that is lower than the mean free path of the photons. A long distance is one that is greater than the mean free path, if still less than the diffusion length. On the smaller scale, perturbations are damped almost instantaneously. On the larger scale, anisotropies are decreased more slowly, with significant degradation happening within one unit of Hubble time.

The Silk Scale and the Silk Mass

Diffusion damping exponentially decreases anisotropies in the CMB on a scale (the Silk scale) much smaller than a degree, or smaller than approximately 3 megaparsecs. This angular scale corresponds to a multipole moment $l \gtrsim 800$. The mass contained within the Silk scale is the *silk mass*. Numerical evaluations of the Silk mass yield results on the order of 10^{13} solar masses at recombination and on the order of the mass of a present-day galaxy or galaxy cluster in the current era:

$$M_S \approx \frac{m_p t_{rec}^{3/2}}{\sqrt{n_{rec}\sigma^3}}$$

Scientists say diffusion damping affects *small* angles and corresponding anisotropies. Other effects operate on a scale called *intermediate* $10 \lesssim l \lesssim 100$ or *large* $l \lesssim 10$. Searches for anisotropies on a small scale are not as difficult as those on larger scales, partly because they may employ ground-based telescopes and their results can be more easily predicted by current theoretical models.

Galaxy Formation

Scientists study photon diffusion damping (and CMB anisotropies in general) because of the insight the subject provides into the question, "How did the universe come to be?". Specifically, primordial anisotropies in the temperature and density of the universe are supposed to be the causes of later large-scale structure formation. Thus it was the amplification of small perturbations in the pre-recombination universe that grew into the galaxies and galaxy clusters of the present era. Diffusion damping made the universe isotropic within distances on the order of the Silk Scale. That this scale corresponds to the size of observed galaxies (when the passage of time is taken into account) implies that diffusion damping is responsible for limiting the size of these galaxies. The theory is that clumps of matter in the early universe became the galaxies that we see today, and the size of these galaxies is related to the temperature and density of the clumps.

Diffusion may also have had a significant effect on the evolution of primordial cosmic magnetic fields, fields which may have been amplified over time to become galactic magnetic fields. However, these cosmic magnetic fields may have been damped by radiative diffusion: just as acoustic oscillations in the plasma were damped by the diffusion of photons, so were magnetosonic waves (waves of ions travelling through a magnetised plasma). This process began before the era of neutrino decoupling and ended at the time of recombination.

DYNAMO THEORY

In geophysics, dynamo theory proposes a mechanism by which a celestial body such as the Earth or a star generates a magnetic field. The theory describes the process through which a rotating, convecting, and electrically conducting fluid can maintain a magnetic field over astronomical time scales.

Dynamo theory describes the process through which a rotating, convecting, and electrically conducting fluid acts to maintain a magnetic field. This theory is used to explain the presence of anomalously long-lived magnetic fields in astrophysical bodies. The conductive fluid in the geodynamo is liquid iron in the outer core, and in the solar dynamo is ionized gas at the tachocline. Dynamo theory of astrophysical bodies uses magnetohydrodynamic equations to investigate how the fluid can continuously regenerate the magnetic field.

It was actually once believed that the dipole, which comprises much of the Earth's magnetic field and is misaligned along the rotation axis by 11.3 degrees, was caused by permanent magnetization of the materials in the earth. This means that dynamo theory was originally used to explain the Sun's magnetic field in its relationship with that of the Earth. However, this theory, which was initially proposed by Joseph Larmor in 1919, has been modified due to extensive studies of magnetic secular variation, paleomagnetism (including polarity reversals), seismology, and the solar system's abundance of elements. Also, the application of the theories of Carl Friedrich Gauss to magnetic observations showed that Earth's magnetic field had an internal, rather than external, origin. There are three requisites for a dynamo to operate:

- An electrically conductive fluid medium.

- Kinetic energy provided by planetary rotation.

- An internal energy source to drive convective motions within the fluid.

In the case of the Earth, the magnetic field is induced and constantly maintained by the convection of liquid iron in the outer core. A requirement for the induction of field is a rotating fluid. Rotation in the outer core is supplied by the Coriolis effect caused by the rotation of the Earth. The coriolis force tends to organize fluid motions and electric currents into columns aligned with the rotation axis. Induction or creation of magnetic field is described by the induction equation,

$$\frac{\partial \mathbf{B}}{\partial t} = \eta \nabla^2 \mathbf{B} + \nabla \times (\mathbf{u} \times \mathbf{B})$$

where u is velocity, B is magnetic field, t is time, and $\eta = 1/\sigma\mu$ is the magnetic diffusivity with electrical conductivity and μ permeability. The ratio of the second term on the right hand side to the first term gives the Magnetic Reynolds number, a dimensionless ratio of advection of magnetic field to diffusion.

Kinematic Dynamo Theory

In kinematic dynamo theory the velocity field is prescribed, instead of being a dynamic variable. This method cannot provide the time variable behavior of a fully nonlinear chaotic dynamo but is useful in studying how magnetic field strength varies with the flow structure and speed.

Using Maxwell's equations simultaneously with the curl of Ohm's Law, one can derive what is basically the linear eigenvalue equation for magnetic fields (B) which can be done when assuming that the magnetic field is independent from the velocity field. One arrives at a critical *magnetic Reynolds number* above which the flow strength is sufficient to amplify the imposed magnetic field, and below which it decays.

The most functional feature of kinematic dynamo theory is that it can be used to test whether a velocity field is or is not capable of dynamo action. By applying a certain velocity field to a small magnetic field, it can be determined through observation whether the magnetic field tends to grow or not in reaction to the applied flow. If the magnetic field does grow, then the system is either capable of dynamo action or is a dynamo, but if the magnetic field does not grow, then it is simply referred to as non-dynamo.

The membrane paradigm is a way of looking at black holes that allows for the material near their surfaces to be expressed in the language of dynamo theory.

Nonlinear Dynamo Theory

The kinematic approximation becomes invalid when the magnetic field becomes strong enough to affect the fluid motions. In that case the velocity field becomes affected by the Lorentz force, and so the induction equation is no longer linear in the magnetic field. In most cases this leads to a quenching of the amplitude of the dynamo. Such dynamos are sometimes also referred to as hydromagnetic dynamos. Virtually all dynamos in astrophysics and geophysics are hydromagnetic dynamos.

Numerical models are used to simulate fully nonlinear dynamos. A minimum of 5 equations are needed. They are as follows. The induction equation, Maxwell's equation:

$$\nabla \cdot B = 0$$

The (sometimes) Boussinesq conservation of mass:

$$\nabla \cdot u = 0$$

The (sometimes) Boussinesq conservation of momentum, also known as the

Navier-Stokes equation:

$$\frac{Du}{Dt} = -\nabla_P + v\nabla^2 u + \rho'g + 2\Omega \times u + \Omega \times \Omega \times R + J \times B$$

where v is the kinematic viscosity, ρ' is the density perturbation that provides buoyancy (for thermal convection $\rho' = \alpha\Delta T$. Ω is the rotation rate of the Earth, and J is the electrical current density. Finally, a transport equation, usually of heat (sometimes of light element concentration):

$$\frac{\partial T}{\partial t} = k\nabla^2 T + \in$$

where T is the temperature $k = k / \rho c_p$ thermal diffusivity with k thermal conductivity, c_p heat capacity, and ρ density, and $\in$ is an optional heat source. Often the pressure is the dynamic pressure, with the hydrostatic pressure and centripetal potential removed. These equations are then non-dimensionalized, introducing the non-dimensional parameters,

$$Ra = \frac{g\alpha TD^3}{vk}, E = \frac{v}{\Omega D^2}, \Pr = \frac{v}{k}, Pm = \frac{v}{\eta}$$

where Ra is the Rayleigh number, E the Ekman number, Pr and Pm the Prandtl and magnetic Prandtl number. Magnetic field scaling is often in Elsasser number units $B = \rho\Omega / \sigma$.

PLASMA ARC CUTTING

Plasma arc cutting, also referred to as plasma fusion cutting or plasma cutting, is a fabrication process which employs superheated, ionized gas funneled through a plasma torch to heat, melt and, ultimately, cut electrically conductive material into custom shapes and designs. This process is suitable for a wide range of metal materials, including structural steel, alloy steel, aluminum, and copper, and can cut through material thicknesses ranging between 0.5 mm to 180 mm.

The plasma cutting process is often presented as an alternative solution to laser cutting, waterjet cutting, and oxy-fuel cutting, and offers certain advantages over these options, including faster cutting times and lower initial investment and operational costs. While plasma cutting demonstrates some advantages over these other cutting processes, its use in some manufacturing applications can be problematic, such as cutting non-conductive material.

Plasma Arc Cutting Process

The plasma arc cutting process is a thermal-based fabrication process which utilizes a constricted, transferred plasma arc to cut through a wide range of metals, including structural steel, alloy steel, aluminum, and copper. While there are several variations available, the basic principles of the process and the necessary components remain the same throughout all of them. The primary plasma arc cutting process includes the following phases:

- Pilot arc initiation.

- Main arc generation.

- Localized heating and melting.

- Material ejection.

- Arc movement.

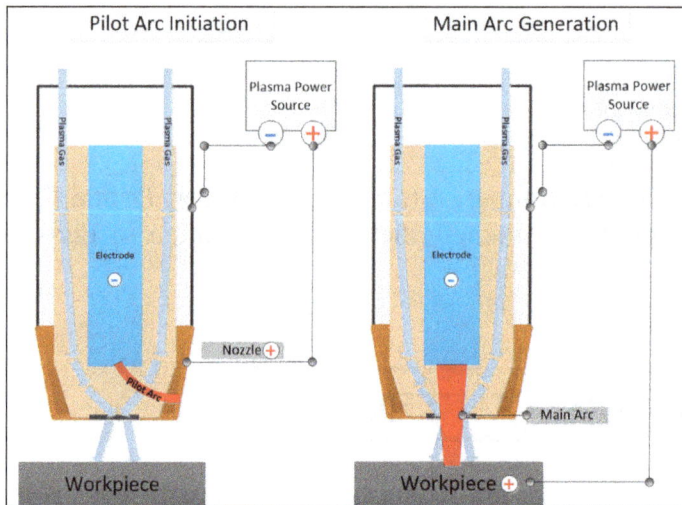

Plasma Arc Electrical Polarities.

Pilot Arc Initiation

The process begins with a start command prompting the power source to generate up to 400VDC of open-circuit voltage—i.e., no-load voltage—and initiating the flow of compressed plasma gas into the plasma torch assembly, which contains an electrode and a plasma nozzle. As illustrated in the *Pilot Arc Initiation* diagram in figure above, the power source also applies a negative voltage to the electrode—establishing it as the cathode of the pilot arc circuit—and closes the normally-open nozzle circuit contacts, placing a temporary, positive voltage on the nozzle—which then serves as the anode of the pilot arc circuit. The arc starting console (ACS) then creates a high-frequency, high-voltage potential between the electrode and nozzle which generates a high-frequency spark. The spark ionizes the plasma gas, allowing it to become electrically conductive and producing a low-resistance current path between the electrode and nozzle. Along this current path, an initial, low-energy arc—i.e., the pilot arc—forms as energy flows and discharges between the two components.

Main Arc Generation

Once initiated, the pilot arc flows out with the plasma gas through the nozzle opening towards the grounded, electrically conductive workpiece, which partially ionizes the area between and forms a new, low-resistance current path. As the gas flow forces the pilot arc to protrude further through the opening, the arc eventually comes into contact with and transfers to the workpiece. As illustrated in the *Main Arc Generation* diagram in figure above, this arc transfer produces the main arc—i.e., the plasma arc which executes the actual cutting operation—and establishes the workpiece as part of the newly-created main arc circuit along with the electrode. The arc transfer also prompts the power source to re-open the normally open nozzle contacts, removing the nozzle from the pilot arc circuit, and the main arc to increase to optimal cutting amperage.

Localized Heating and Melting

The nozzle constricts the ionized gas and the main arc as they flow through the nozzle opening, increasing the plasma's energy density and velocity. Plasma cutting machines produce plasma up to 20,000°C which moves towards the workpiece at up to three times the speed of sound. This thermal and kinetic energy is used for the cutting operation.

The plasma arc cutting process employs a melt-and-blow cutting method, which heats, melts and vaporizes a localized area of the workpiece. As the plasma strikes the workpiece's surface, the workpiece material absorbs the thermal energy of the arc and plasma gas, increasing the internal energy and generating heat, which weakens the material and allows it to be removed to produce the desired cuts.

Material Ejection

The weakened material of the workpiece is expelled out of the kerf—the width of material removed and of the cut product—by the kinetic energy of the plasma gas employed. The optimal flow of the plasma gas is determined by the current and the nozzle, with too low or too high plasma flow levels leading to less precise cuts and component failure.

Arc Movement

Once the localized heating, melting, and vaporizing of the workpiece have started, the plasma arc is manually or automatically moved across the workpiece's surface to produce the full cut. In the case of handheld plasma arc cutting systems, the operator manually initiates the process and moves the torch across the surface to create the desired cuts. For automated plasma arc cutting systems, the machine is programmed to move the torch head at the optimal speed to ensure precise and accurate cuts.

Handheld plasma arc cutting torch
performing cutting operation.

Variants of the Plasma Arc Cutting Process

The basic principles behind the plasma arc cutting process remain the same across the different variations available. However, each process variation provides particular advantages in regards to manufacturing applications based on the material being cut and its properties, the power output, and specific requirements of the application. The variations are typically differentiated based on their cooling system, the type of plasma gas, the design of the electrode, and the type of plasma employed.

Some of the plasma arc cutting options available include:

- Standard (or conventional) plasma arc cutting.

- Plasma arc cutting utilizing a secondary medium.

- Water injection plasma arc cutting.

Standard (or Conventional) Plasma Arc Cutting

In conventional plasma arc cutting, the plasma cutting equipment includes the plasma torch assembly, which utilizes a single plasma gas serving as both the ignition gas and the cutting gas. Typically, the standard process employs nitrogen, oxygen, or a hydrogen-argon mixture. Constriction of the plasma arc and gas is performed only by the nozzle without the aid of any secondary medium. Water or air can be used as the coolant for the plasma torch.

Plasma Arc Cutting with Secondary Medium

For plasma arc cutting which employs a secondary medium, an additional medium—i.e., water or gas—is siphoned into the plasma torch to constrict the plasma arc further and to produce specific characteristics for the particular cutting application.

Introducing a secondary gas to the plasma arc cutting process can increase power density, the quality of the cut, and cutting speed. Additionally, the secondary gas can reduce damage to the system and the risk of double arcing, and help extend the lifespan of the consumable torch parts. This type of plasma arc cutting is suitable for metal plates in thicknesses up to 75 mm.

Some of the more common secondary gas combinations include:

- Air, oxygen, and nitrogen for steel cutting.

- Nitrogen, argon-H_2, and CO_2 for stainless steel cutting.

- Argon-H_2, nitrogen, and CO_2 for aluminum cutting.

Introducing water to the plasma arc cutting process can produce workpiece surfaces with higher reflectivity. The water acts as a barrier or shield during the cutting process as it is siphoned into the plasma torch, discharged, and evaporated by the plasma arc. This type of plasma arc cutting is suitable for aluminum and high-alloy steels in thicknesses up to 50 mm.

Water Injection Plasma Arc Cutting

Water injection plasma arc cutting also employs water during the cutting process. Water is injected into the plasma torch, which further constricts the plasma arc. As opposed to plasma arc cutting with a secondary medium, the majority of the water remains unevaporated and instead acts as a coolant for the plasma torch components and the workpiece. The cooling effect of the water allows for less material distortion, higher quality cuts, and extended lifespan for consumable torch parts. This type of plasma arc cutting is suitable for use with underwater plasma cutting machines, and metals in thicknesses ranging from 3 mm to 75 mm.

Other Variations

Other variations of plasma arc cutting include:

- Plasma arc cutting with increased constriction.

- Underwater plasma arc cutting.

- Plasma gouging.

- Plasma marking.

Separate from the previously mentioned variations, increased *plasma arc constriction* is achieved by using specialized nozzles which allow for particular capabilities, such as rotating the plasma gas or adjusting the nozzle during the cutting process.

Underwater plasma arc cutting is performed between 60 mm to 100 mm below water, which allows for noise, dust, and air pollution reduction, but requires more energy and more cutting time than atmospheric plasma arc cutting.

Plasma gouging and plasma marking are processes that do not typically cut through the workpiece; plasma gouging removes just the surface material of the workpiece to produce a smoother surface, while plasma marking produces surface marks on finished components.

The plasma arc cutting process offers a variety of options which can suit a wide range of manufacturing applications. The suitability of each variation depends on the specifications and requirements of the cutting application.

Plasma Arc Cutting Machine

While the laser cutting process utilizes laser cutting machines and the waterjet cutting process/serviceutilizes waterjet cutting machinery with pressurized water and abrasives, the plasma arc cutting process employs plasma cutting equipment to produce the desired cuts on the workpiece. Plasma cutting machines vary from model to model and application and application with setups ranging from the simple (e.g., handheld torches attached to a power source) to the complex (e.g., programmable and automated CNC machinery). The basic equipment setup for these plasma cutting machines includes a plasma power source, arc starting console, torch assembly, gas supply, cooling system, and an electrically conductive workpiece.

- Power Source: The power source provides the energy to initiate the pilot arc and maintain the main arc throughout the plasma arc cutting process. They typically have high, non-load voltages (i.e., open-circuit voltages) ranging from 240VDC to 400VDC to produce the pilot arc, but only require 50VDC to 60VDC to sustain the main arc once it is produced.

- Arc Starting Console (ACS): The ACS produces the initial spark which initiates the pilot arc circuit.

- Gases and Media Employed: Plasma gases are categorized into ignition gases (ignites the plasma arc), cutting gases (used with the plasma arc during the cutting process), and secondary gases (constricts and cools the plasma arc). Gases employed can be inert, reactive, or a mixture of the two previous types. Water is also employed as a secondary medium during the cutting process.

- Torch Assembly: The torch assembly and parts includes the electrode and the nozzle, is connected to the power supply, and utilizes the plasma and cutting gases to initiate and perform the plasma cutting operation.

- Cooling System: The cooling system cools the torch assembly components and workpiece, extending the lifespan of the consumables. The system can be either water or gas-cooled.

- Workpiece: The workpiece is the material being cut. The material must be electrically conductive to be plasma cut as the workpiece serves as a component of the main arc circuit.

Consumable plasma torch nozzles.

Other options for plasma arc cutting setups include cutting benches, air pollution control equipment, and overhead track systems. The cutting bench serves as a work surface for cutting the workpiece, and the control equipment as a means of removing emissions formed during the cutting process. For automated cutting machines, the torch is suspended overhead on a track system to allow for movement across the workpiece's surface.

Material Considerations

As the plasma arc cutting process employs transferred plasma arcs, its use is limited to cutting only materials that are electrically conductive. However, it is suitable for a wide range of metals, including:

- Structural steel.

- Non-alloy, low-ally, and high-alloy steel.

- Aluminum.

- Clad metal plates.

Plasma arc cutting can also be used on materials such as copper, brass, titanium, and cast iron, although some of their melting temperatures may prove to be problematic in achieving a high-quality edge cut. Depending on the specifications of the plasma arc cutting machine and the workpiece material, the process is capable of cutting through material thicknesses ranging between 0.5 mm to 180 mm.

Alternative Cutting Processes

Some of the advantages of plasma arc cutting demonstrated over other cutting methods include:

- Faster turnaround time.

- Higher quality cuts.

- Capabilities for handling thicker materials.

- Minimized risk of material warping.

- Lower equipment and operational costs.

Despite these advantages, however, it may not be appropriate for every manufacturing application, and other cutting processes may prove more suitable and cost-effective. Alternatives to plasma arc cutting include oxy-fuel cutting, waterjet cutting, and laser cutting.

References

- Dusty-plasma: plasma-universe.com, Retrieved 23 January, 2020

- "Study sheds light on turbulence in astrophysical plasmas : Theoretical analysis uncovers new mechanisms in plasma turbulence". MIT News. Retrieved 2018-02-20

- Oppenheim, Meers M.; Endt, Axel F. Vom; Dyrud, Lars P. (October 2000). "Electrodynamics of meteor trail evolution in the equatorial E-region ionosphere". Geophysical Research Letters. 27 (19): 3173. Doi:10.1029/1999GL000013

- Double-layer: plasma-universe.com, Retrieved 24 February, 2020

- Lüttgens, Günter; Lüttgens, Sylvia; Schubert, Wolfgang (2017). Static Electricity: Understanding, Controlling, Applying. John Wiley and Sons. P. 94. ISBN 978-3527341283

- Dynamo-theory, Earth-dynamo, Sources, EPS281r, Courses, eli, climate: courses.seas.harvard.edu, Retrieved 25 March, 2020

- "Simple but challenging: The Universe according to Planck". The European Space Agency (ESA). 2013-03-21. Retrieved 2013-05-01

- Understanding-plasma-arc-cutting, custom-manufacturing-fabricating: thomasnet.com, Retrieved 26 April, 2020

- Johnson W.R.; Nielsen J.; Cheng K.T. (2012). "Thomson scattering in the average-atom approximation". Physical Review. 86 (3): 036410. Arxiv:1207.0178. Bibcode:2012phrve..86c6410j. Doi:10.1103/physreve.86.036410. PMID 23031036

Waves in Plasmas

Waves in plasmas refer to an interconnected set of particles and fields that propagate periodically. Electromagnetic electron wave, ion acoustic wave, and Alfvén wave are some of the examples of these waves. Plasma oscillation, upper and lower hybrid oscillation are also studied under it. The topics elaborated in this chapter will help in gaining a better perspective about these waves in plasma.

A plasma is an ionized gas consisting of charged particles (e.g., electrons and ions). Various waves can be excited easily in a plasma. Wave phenomena have been an important subject in the plasma research community.

The plasma is nearly charge neutral. So the $\nabla \cdot \vec{E} = 0$ still holds. However, neither the conduction current nor the displacement current can be ignored. Waves in plasma is different from the waves in vacuum and in conductors.

Effective Permittivity in Plasma

In vacuum, the phase velocity of an EM wave is:

$$c = \frac{\omega}{k} = \frac{1}{\sqrt{\epsilon_0 \, \mu_0}}$$

This can be generalized for phase velocity of waves in matter:

$$v = \frac{\omega}{k} = \frac{1}{\sqrt{\epsilon \, \mu}}$$

In plasma $\mu = \mu_0$, but $\epsilon \neq \epsilon_0$

Electric field, magnetic field, and generally any quantity in plasma can be divided into two parts: DC part that does not depend on time and parts that associated with waves:

$$Q_{total} = Q_{DC} + Q(\vec{r}, t)$$

We study now only the part associated with waves, Q, and assume,

$$Q = Q_{0e}{}^{i(\vec{k}\cdot\vec{r} - \omega t)}$$

In plasma, the current is predominately carried by electrons, as in a conductor, because the mass of a electron is small compared with that of ions. The electrons in plasma experience the electric force and suffer from collision with ions. The velocity of electrons has been derived before (when we study waves in conductors):

$$\vec{v} = -\frac{e}{m(v - i\omega)} \vec{E}$$

$$\vec{J} = -\frac{ne^2}{m(v - i\omega)} \vec{E}$$

If there is no DC magnetic field (So we can ignore $\vec{v} \times \vec{B}$ term in the equation of motion).

In most cases, the electron density, and thus the collision frequency v, are much smaller than in conductors. So:

$$\vec{j} \simeq -\frac{ne^2}{mi\omega} \vec{E} = i\frac{ne^2}{m\omega} \vec{E}$$

The 4th Maxwell's equation becomes:

$$\nabla \times \vec{B} = \mu_0 \left(i \frac{ne^2}{m\omega} \vec{E} = \epsilon_0 \frac{\partial \vec{E}}{\partial t} \right)$$

$$= \mu_0 \left(i \frac{ne^2}{m\omega} \vec{E} = i\omega \epsilon_0 \vec{E} \frac{\partial \vec{E}}{\partial t} \right)$$

$$= i\omega \epsilon_0 \mu_0 \left(1 - \frac{ne^2}{m \epsilon_0 \omega^2} \right) \vec{E}$$

Let,

$$\omega_p = \sqrt{\frac{ne^2}{m \epsilon_0}}$$

and recall $\vec{H} = \vec{B} / \mu_0$, we find,

$$\nabla \times \vec{H} = -i\omega \epsilon_0 \left(1 - \frac{\omega_p^2}{\omega^2} \right),$$

For low frequency wave, $\omega_p \gg \omega$, conduction current dominates (like waves in a conductor). For high frequency wave, $\omega_p \ll \omega$, displacement current dominates (like waves in vacuum). If we define an effective permittivity in plasma,

$$\epsilon(\omega) = \epsilon_0 \left(1 - \frac{\omega_p^2}{\omega^2} \right)$$

The 4th Maxwell's equation for a monochromatic plane wave in plasma becomes,

$$\nabla \times \vec{H} = -i\omega \in \vec{E} = \in \frac{\partial \vec{E}}{\partial t} \ ,$$

similar to the equation for a monochromatic plane wave in the vacuum, except $\varepsilon_o \rightarrow \varepsilon$. Therefore, all the results for waves in vacuum can be used in plasma wave with the modification $\varepsilon_o \rightarrow \varepsilon$. Note that the effective permittivity in plasma depends on the frequency ω. Note also $\in < \in_0$.

Dispersion Relation

The phase velocity of plasma waves:

$$v = \frac{\omega}{k} = \frac{1}{\sqrt{\mu \in}}$$

$$= \frac{1}{\sqrt{\mu_0 \in_0}} \cdot \frac{1}{\sqrt{1 - \frac{\omega_p^2}{\omega^2}}}$$

$$= \frac{c}{\sqrt{1 - \frac{\omega_p^2}{\omega^2}}} > c$$

In plasma, $\in$ depends on ω and thus the phase velocity depends also on ω. The wave in this case is dispersive.

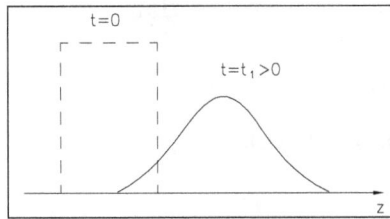

A square wave contains the fundamental frequency and its higher harmonics. If the phase velocity depends on frequency, it will spread out while it propagates.

The dependence of ω on k is called dispersion relation.

We can solve eq. above for ω,

$$\omega^2 = \frac{c^2}{1 - \frac{\omega_p^2}{\omega^2}}$$

$$\frac{\omega^2}{k^2} = \left(1 - \frac{\omega_p^2}{\omega^2}\right) = C^2 k^2$$

$$\underline{\omega^2 = \omega_p^2 + (ck)^2}$$ dispersion relation in plasma

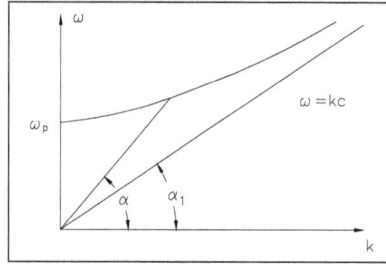

The dispersion relation of EM waves in vacuum is a straight line (non-dispersive) with a slope $\tan \alpha = c$.

The dispersion relation of EM waves in plasma is above the line $\omega = ck$ because,

$$\omega = \sqrt{\omega_p^2 + (ck)^2} \geq ck,$$

but approaches the line $\omega = ck$ when k becomes large because,

$$\omega = \sqrt{\omega_p^2 + (ck)^2} \simeq ck, \text{ When } k \gg \frac{\omega_p}{c}.$$

Propagation and Reflection of EM Waves in Plasma

Assume a plane wave propagating in +z direction,

$$\vec{E} = \vec{E}_{0e}{}^{i(kz-\omega t)}$$

From the dispersion relation, we obtain,

$$k = \frac{1}{c}\sqrt{\omega^2 - \omega_p^2}$$

- If $\omega > \omega_p$ (like in vacuum) k is real $\vec{E} = \vec{E}0e^{i(kz-\omega t)}$, wave propagates without decay.

- If $\omega < \omega_p$ (like in conductor)

$$k = i\frac{1}{c}\sqrt{\omega_p^2 - \omega^2} = i|k| \text{ is imaginary.}$$

$$\vec{E} = \vec{E}0e^{-i\omega t}e^{-|k|z}$$

Waves decays in z direction and will be reflected.

If $\omega \ll \omega_p$, $k \simeq i\dfrac{\omega p}{c} = i\dfrac{1}{\delta}$

$\quad \delta = \dfrac{c}{\omega_p}$ skin depth in plasma.

Short Wave Communication

Ionospheric plasma (height: 50 km to 100 km) has a typical density of $10^{13}/m^3$.

$$f_p = \frac{\omega_p}{2\pi} = \frac{1}{2\pi}\sqrt{\frac{ne2}{m\,\epsilon_0}} = 28\text{ MHz}$$

Short wave radio (f ~ 10 MHz) relies on the multiple reflection between the ionospheric plasma layer and the earth to reach a distant receiver.

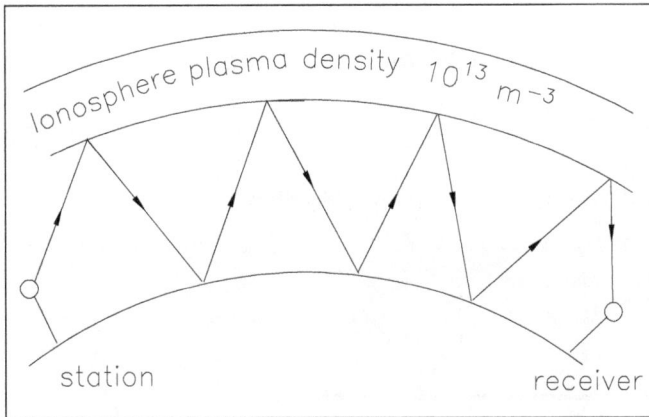

Earth is "conductor" ($\sigma \sim 10^{-2}\ S/m$) as long as the impedance,

$$|Z| = \left|\sqrt{\frac{i\omega\mu0}{\sigma}}\right| \ll Z_{air} = \sqrt{\frac{\mu0}{\epsilon_0}}$$

which requires,

$$\omega \ll \frac{\sigma}{\epsilon_0}$$

or

$$f \ll \frac{\sigma}{2\pi\,\epsilon_0} = \frac{10^{-2}}{2\pi\times 8.85\times 10^{-12}} = 180\,\text{MH}_z$$

For $f = 10$ MHz, the earth is good conductor.

Group Velocity

According to Einstein's relativity, nothing should propagate faster than the light speed c.

In plasma, phase velocity,

$$vp = \frac{\omega}{k} = \frac{c}{\sqrt{1 - \frac{\omega_p^2}{\omega^2}}} > c$$

But the phase velocity does not correspond to the information (energy) propagation velocity.

The information is propagating at the group velocity,

$$v_g = \frac{d\omega}{dk}$$

For non-dispersive waves, e.g., EM waves in vacuum,

$$\omega = ck \quad v_p = v_g = c.$$

In plasma, wave is dispersive. The group velocity is,

$$vp = \frac{d}{dk} = c\sqrt{1 - \underline{\quad}} > c\;.$$

PLASMA OSCILLATION

Plasma oscillations, also known as Langmuir waves (after Irving Langmuir), are rapid oscillations of the electron density in conducting media such as plasmas or metals in the ultraviolet region. The oscillations can be described as an instability in the dielectric function of a free electron gas. The frequency only depends weakly on the wavelength of the oscillation. The quasiparticle resulting from the quantization of these oscillations is the plasmon.

Langmuir waves were discovered by American physicists Irving Langmuir and Lewi Tonks in the 1920s. They are parallel in form to Jeans instability waves, which are caused by gravitational instabilities in a static medium.

Mechanism

Consider an electrically neutral plasma in equilibrium, consisting of a gas of positively charged ions and negatively charged electrons. If one displaces by a tiny amount an electron or a group of electrons with respect to the ions, the Coulomb force pulls the electrons back, acting as a restoring force.

Cold Electrons

If the thermal motion of the electrons is ignored, it is possible to show that the charge density oscillates at the *plasma frequency*,

$$\omega_{pe} = \sqrt{\frac{n_e e^2}{m^* \varepsilon_0}}, [\text{rad}/\text{s}] \quad (\text{SI units}),$$

$$\omega_{pe} = \sqrt{\frac{4\pi n_e e^2}{m^*}}, [rad/s] \, (\text{cgs units}),$$

where n_e is the number density of electrons, e is the electric charge, m^* is the effective mass of the electron, and ε_0 is the permittivity of free space. Note that the above formula is derived under the approximation that the ion mass is infinite. This is generally a good approximation, as the electrons are so much lighter than ions. (This expression must be modified in the case of electron-positron plasmas, often encountered in astrophysics). Since the frequency is independent of the wavelength, these oscillations have an infinite phase velocity and zero group velocity.

Note that, when , $m^* = m_e$ the plasma frequency, ω_{pe}, depends only on physical constants and electron density n_e. The numeric expression for angular plasma frequency is:

$$f_{pe} = \frac{\omega_{pe}}{2\pi} [\text{Hz}]$$

Metals are only transparent to light with a frequency higher than the metal's plasma frequency. For typical metals such as aluminium or silver, n_e is approximately 10^{23} cm^{-3}, which brings the plasma frequency into the ultraviolet region. This is why most metals reflect visible light and appear shiny.

Warm Electrons

When the effects of the electron thermal speed $v_{e,th} = \sqrt{\frac{k_B T_e}{m_e}}$ are taken into account, the electron pressure acts as a restoring force as well as the electric field and the oscillations propagate with frequency and wavenumber related by the longitudinal Langmuir wave:

$$\omega^2 = \omega_{pe}^2 + \frac{3k_B T_e}{m_e} k^2 = \omega_{pe}^2 + 3k^2 v_{e,th}^2,$$

called the Bohm-Gross dispersion relation. If the spatial scale is large compared to the

Debye length, the oscillations are only weakly modified by the pressure term, but at small scales the pressure term dominates and the waves become dispersionless with a speed of $\sqrt{3} \cdot v_{e,th}$. For such waves, however, the electron thermal speed is comparable to the phase velocity, i.e.,

$$v \sim v_{ph} \stackrel{def}{=} \frac{\omega}{k},$$

so the plasma waves can accelerate electrons that are moving with speed nearly equal to the phase velocity of the wave. This process often leads to a form of collisionless damping, called Landau damping. Consequently, the large-k portion in the dispersion relation is difficult to observe and seldom of consequence.

In a bounded plasma, fringing electric fields can result in propagation of plasma oscillations, even when the electrons are cold.

In a metal or semiconductor, the effect of the ions' periodic potential must be taken into account. This is usually done by using the electrons' effective mass in place of m.

UPPER HYBRID OSCILLATION

In plasma physics, an upper hybrid oscillation is a mode of oscillation of a magnetized plasma. It consists of a longitudinal motion of the electrons perpendicular to the magnetic field with the dispersion relation,

$$\omega^2 = \omega_{pe}^2 + \omega_{ce}^2 + 3k^2 v_{e,th}^2,$$

where (in cgs units),

$$\omega_{pe} = (4\pi n_e e^2 / m_e)^{1/2}$$

is the electron plasma frequency,

$$\omega_{ce} = eB / m_e c$$

is the electron cyclotron frequency.

This oscillation is closely related to the plasma oscillation found in unmagnetized plasmas or parallel to the magnetic field, where the ω_{pe} term arises from the electrostatic Coulomb restoring force and the $3k^2 v_{e,th}^2$ term arises from the restoring force of electron pressure. In the upper hybrid oscillation there is an additional restoring force due to the Lorentz force. Consider a plane wave where all perturbed quantities vary as exp($i(kx-\omega t)$). If the displacement in the direction of propagation is δ_x, then.

$$v_x = -i\omega\delta$$

$$f_y = nev_x B_z/c = -i\omega(neB_z/c)\delta$$

$$v_y = -f_y/i\omega nm = (eB_z/mc)\delta$$

$$f_x = -nev_y B_z/c = -(nm)(eB_z/mc)^2\delta$$

$$a_x = -\omega_{ce}^2\delta$$

Thus the perpendicular magnetic field effectively provides a harmonic restoring force with a frequency ω_{ce}, explaining the third term in the dispersion relation. The particle orbits (or fluid trajectories) are ellipses in the plane perpendicular to the magnetic field, elongated in the direction of propagation.

The frequency of long wavelength oscillations is a "hybrid", or mix, of the electron plasma and electron cyclotron frequencies,

$$\omega_h^2 = \omega_{pe}^2 + \omega_{ce}^2,$$

and is known as the upper hybrid frequency. There are also a lower hybrid frequency and lower hybrid oscillations.

For propagation at angles oblique to the magnetic field, two modes exist simultaneously. If the plasma frequency is higher than the cyclotron frequency, then the upper hybrid oscillation transforms continuously into the plasma oscillation. The frequency of the other mode varies between the cyclotron frequency and zero. Otherwise, the frequency of the mode related to the upper hybrid oscillation remains above the cyclotron frequency, and the mode related to the plasma oscillation remains below the plasma frequency. In particular, the frequencies are given by,

$$\omega^2 = (1/2)\omega_h^2\left(1\pm\sqrt{1-\left(\frac{\cos\theta}{\omega_h^2/2\omega_c\omega_p}\right)^2}\right)$$

LOWER HYBRID OSCILLATION

In plasma physics, a lower hybrid oscillation is a longitudinal oscillation of ions and electrons in a magnetized plasma. The direction of propagation must be very nearly perpendicular to the stationary magnetic field, within about $\sqrt{m_e/m_i}$ radians. Otherwise the electrons can move along the field lines fast enough to shield the oscillations in potential. The frequency of oscillation is,

$$\omega = [(\Omega_i\Omega_e)^{-1} + \omega_{pi}^{-2}]^{-1/2}$$

where Ω_i is the ion cyclotron frequency, Ω_e is the electron cyclotron frequency and ω_{pi} is the ion plasma frequency. This is the lower hybrid frequency, so called because it is a "hybrid", or mixture, of two frequencies. There are also an upper hybrid frequency and upper hybrid oscillation.

The lower hybrid oscillation is unusual in that the ion and electron masses play an equally important role. This mode is relatively unimportant in practice because the necessary precise orientation relative to the magnetic field is seldom achieved. Exceptions are the use of lower hybrid waves to heat and drive current in fusion plasmas, and the lower hybrid drift instability, which was thought to be an important determinant of transport in the Field-Reversed Configuration.

APPLETON–HARTREE EQUATION

The Appleton–Hartree equation, sometimes also referred to as the Appleton–Lassen equation is a mathematical expression that describes the refractive index for electromagnetic wave propagation in a cold magnetized plasma. The Appleton–Hartree equation was developed independently by several different scientists, including Edward Victor Appleton, Douglas Hartree and German radio physicist H. K. Lassen. Lassen's work, completed two years prior to Appleton and five years prior to Hartree, included a more thorough treatment of collisional plasma; but, published only in German, it has not been widely read in the English speaking world of radio physics. Further, regarding the derivation by Appleton, it was noted in the historical study by Gilmore that Wilhelm Altar (while working with Appleton) first calculated the dispersion relation in 1926.

Equation:

The dispersion relation can be written as an expression for the frequency (squared), but it is also common to write it as an expression for the index of refraction $n^2 = \left(\dfrac{ck}{\omega}\right)^2$.

The equation is typically given as follows:

$$n^2 = 1 - \frac{X}{1 - iZ - \dfrac{\frac{1}{2}Y^2\sin^2\theta}{1 - X - iZ} \pm \dfrac{1}{1 - X - iZ}\left(\frac{1}{4}Y^4\sin^4\theta + Y^2\cos^2\theta(1 - X - iZ)^2\right)^{1/2}}$$

or, alternatively, with damping term Z = 0 and rearranging terms:

$$n^2 = 1 - \frac{X(1 - X)}{1 - X - \frac{1}{2}Y^2\sin^2\theta \pm \left(\left(\frac{1}{2}Y^2\sin^2\theta\right)^2 + (1 - X)^2 Y^2\cos^2\theta\right)^{1/2}}$$

Definition of Terms

n = complex refractive index.

$$i = \sqrt{-1}$$

$$X = \frac{\omega_0^2}{\omega^2}$$

$$Y = \frac{\omega_H}{\omega}$$

$$Z = \frac{v}{\omega}$$

v = electron collision frequency.

$\omega = 2\pi f$ (radial frequency).

f = wave frequency (cycles per second, or Hertz).

$$\omega_0 = 2\pi f_0 = \sqrt{\frac{Ne^2}{\epsilon_0 m}} = \text{electron plasma frequency.}$$

$$\omega_H = 2\pi f_H = \frac{B_0 |e|}{m} = \text{electron gyro frequency.}$$

ϵ_0 = permittivity of free space.

B_0 = ambient magnetic field strength.

e = electron charge.

m = electron mass.

θ = angle between the ambient magnetic field vector and the wave vector.

Modes of Propagation

The presence of the $\pm$ sign in the Appleton–Hartree equation gives two separate solutions for the refractive index. For propagation perpendicular to the magnetic field, i.e., $\mathbf{k} \parallel \mathbf{B}_0$, the '+' sign represents the "ordinary mode," and the '−' sign represents the "extraordinary mode." For propagation parallel to the magnetic field, i.e., $\mathbf{k} \perp \mathbf{B}_0$, the '+' sign represents a left-hand circularly polarized mode, and the '−' sign represents a right-hand circularly polarized mode.

$\mathbf{k}$ is the vector of the propagation plane.

Reduced Forms

Propagation in a Collisionless Plasma

If the electron collision frequency v is negligible compared to the wave frequency of interest ω, the plasma can be said to be "collisionless." That is, given the condition.

$$v \ll \omega,$$

we have

$$Z = \frac{v}{\omega} \ll 1,$$

so we can neglect the z terms in the equation. The Appleton–Hartree equation for a cold, collisionless plasma is therefore,

$$n^2 = 1 - \frac{X}{1 - \dfrac{\frac{1}{2} Y^2 \sin^2 \theta}{1 - X} \pm \dfrac{1}{1 - X} \left(\frac{1}{4} Y^4 \sin^4 \theta + Y^2 \cos^2 \theta (1 - X)^2 \right)^{1/2}}$$

Quasi-longitudinal Propagation in a Collisionless Plasma

If we further assume that the wave propagation is primarily in the direction of the magnetic field, i.e., $\theta \approx 0$, we can neglect the $Y^4 \sin^4 \theta$ term above. Thus, for quasi-longitudinal propagation in a cold, collisionless plasma, the Appleton–Hartree equation becomes,

$$n^2 = 1 - \frac{X}{1 - \dfrac{\frac{1}{2} Y^2 \sin^2 \theta}{1 - X} \pm Y \cos \theta}$$

ELECTROMAGNETIC ELECTRON WAVE

In plasma physics, an electromagnetic electron wave is a wave in a plasma which has a magnetic field component and in which primarily the electrons oscillate.

In an unmagnetized plasma, an electromagnetic electron wave is simply a light wave modified by the plasma. In a magnetized plasma, there are two modes perpendicular to the field, the O and X modes, and two modes parallel to the field, the R and L waves.

Cut-off Frequency and Critical Density

In an unmagnetized plasma for the high frequency or low electron density limit, i.e. for or $\omega \gg \omega_{pe} = (n_e e^2 / m_e \epsilon_0)^{1/2}$ or $n_e \ll m_e \omega^2 \epsilon_0 / e^2$ where ω_{pe} is the plasma frequency, the wave speed is the speed of light in vacuum. As the electron density increases, the phase velocity increases and the group velocity decreases until the cut-off frequency where the light frequency is equal to ω_{pe}. This density is known as the critical density for the angular frequency ω of that wave and is given by:

$$n_c = \frac{\varepsilon_o m_e}{e^2} \omega^2 \text{ (SI units)}$$

If the critical density is exceeded, the plasma is called over-dense.

In a magnetized plasma, except for the O wave, the cut-off relationships are more complex.

O Wave

The O wave is the *ordinary* wave in the sense that its dispersion relation is the same as that in an unmagnetized plasma. It is plane polarized with $E_1 \parallel B_0$. It has a cut-off at the plasma frequency.

X Wave

The X wave is the "extraordinary" wave because it has a more complicated dispersion relation. It is partly transverse (with $E_1 \perp B_0$) and partly longitudinal. As the density is increased, the phase velocity rises from c until the cut-off at ω_R is reached. As the density is further increased, the wave is evanescent until the resonance at the upper hybrid frequency ω_h. Then it can propagate again until the second cut-off at ω_L. The cut-off frequencies are given by:

$$\omega R = \frac{1}{2}\left[\omega_c + \left(\omega_c^2 + 4\omega_p^2 \right)^{\frac{1}{2}} \right]$$

$$\omega L = \frac{1}{2}\left[-\omega_c + \left(\omega_c^2 + 4\omega_p^2 \right)^{\frac{1}{2}} \right]$$

where ω_c is the electron cyclotron resonance frequency, and ω_p is the electron plasma frequency.

R Wave and L Wave

The R wave and the L wave are right-hand and left-hand circularly polarized, respectively. The R wave has a cut-off at ω_R (hence the designation of this frequency) and a resonance at ω_c. The L wave has a cut-off at ω_L and no resonance. R waves at frequencies below $\omega_c/2$ are also known as whistler modes.

Dispersion Relations

The dispersion relation can be written as an expression for the frequency (squared), but it is also common to write it as an expression for the index of refraction ck/ω (squared).

Summary of electromagnetic electron waves		
Conditions	Dispersion relation	Name
$\vec{B}_0 = 0$	$\omega^2 = \omega_p^2 + k^2 c^2$	Light wave
$\vec{k} \perp \vec{B}_0, \vec{E}_1 \parallel \vec{B}_0$	$\dfrac{c^2 k^2}{\omega^2} = 1 - \dfrac{\omega_p^2}{\omega^2}$	O wave
$\vec{k} \perp \vec{B}_0, \vec{E}_1 \perp \vec{B}_0$	$\dfrac{c^2 k^2}{\omega^2} = 1 - \dfrac{\omega_p^2}{\omega^2} \dfrac{\omega^2 - \omega_p^2}{\omega^2 - \omega_h^2}$	X wave
$\vec{k} \parallel \vec{B}_0$ (right circ. pol.)	$\dfrac{c^2 k^2}{\omega^2} = 1 - \dfrac{\omega_p^2 / \omega^2}{1 - \omega_c / \omega}$	R wave (whistler mode)
$\vec{k} \parallel \vec{B}_0$ (left circ. pol.)	$\dfrac{c^2 k^2}{\omega^2} = 1 - \dfrac{\omega_p^2 / \omega^2}{1 - \omega_c / \omega}$	L wave

ION ACOUSTIC WAVE

In plasma physics, an ion acoustic wave is one type of longitudinal oscillation of the ions and electrons in a plasma, much like acoustic waves traveling in neutral gas. However, because the waves propagate through positively charged ions, ion acoustic waves can interact with their electromagnetic fields, as well as simple collisions. In plasmas, ion acoustic waves are frequently referred to as acoustic waves or even just sound waves. They commonly govern the evolution of mass density, for instance due to pressure gradients, on time scales longer than the frequency corresponding to the relevant length scale. Ion acoustic waves can occur in an unmagnetized plasma or in a magnetized plasma parallel to the magnetic field. For a single ion species plasma and in the long wavelength limit, the waves are dispersionless $(\omega = v_s k)$ with a speed given by,

$$v_s = \sqrt{\frac{\gamma_e Z K_B T_e + \gamma_i K_B T_i}{M}}$$

where K_B is Boltzmann's constant, M is the mass of the ion, Z is its charge, T_e is the temperature of the electrons andT T_i is the temperature of the ions. Normally γ_e is taken to be unity, on the grounds that the thermal conductivity of electrons is large enough to keep them isothermal on the time scale of ion acoustic waves, and γ_i is taken to be 3, corresponding to one-dimensional motion. In collisionless plasmas, the electrons are often much hotter than the ions, in which case the second term in the numerator can be ignored.

Derivation

We derive the ion acoustic wave dispersion relation for a linearized fluid description of a plasma with electrons and N ion species. We write each quantity as $X = X_0 + \delta \cdot X_1$ where subscript 0 denotes the "zero-order" constant equilibrium value, and 1 denotes the first-order perturbation. δ is an ordering parameter for linearization, and has the physical value 1. To linearize, we balance all terms in each equation of the same order in δ. The terms involving only subscript-0 quantities are all order δ^0 and must balance, and terms with one subscript-1 quantity are all order δ^1 and balance. We treat the electric field as order-1. $\left(\vec{E}_0 = 0 \right)$ and neglect magnetic fields,

Each species s is described by mass m_s, charge $q_s = Z_s e$, number density n_s, flow velocity $\vec{u}_s$, and pressure p_s. We assume the pressure perturbations for each species are a Polytropic process, namely $p_{s1} = \gamma_s T_{s0} n_{s1}$ for species s. To justify this assumption and determine the value of γ_s, one must use a kinetic treatment that solves for the species distribution functions in velocity space. The polytropic assumption essentially replaces the energy equation.

Each species satisfies the continuity equation,

$$\partial_t n_s + \nabla \cdot (n_s \vec{u}_s) = 0$$

and the momentum equation,

$$\partial_t \vec{u}_s + \vec{u}_s \cdot \nabla \vec{u}_s = \frac{Z_s e}{m_s} \vec{E} - \frac{\nabla p_s}{n_s}.$$

We now linearize, and work with order-1 equations. Since we do not work with T_{s1} due to the polytropic assumption (but we do *not* assume it is zero), to alleviate notation we use T_s for T_{s0}. Using the ion continuity equation, the ion momentum equation becomes,

$$(-m_i \partial_{tt} + \gamma_i T_i \nabla^2) n_{i1} = Z_i e n_{i0} \nabla \cdot \vec{E}_1$$

We relate the electric field $\vec{E}_1$ to the electron density by the electron momentum equation:

$$n_{e0} m_e \partial_t \vec{v}_{e1} = -n_{e0} e \vec{E}_1 - \gamma_e T_e \nabla n_{e1},$$

We now neglect the left-hand side, which is due to electron inertia. This is valid for waves with frequencies much less than the electron plasma frequency $(n_{e0}e^2/\epsilon_0 m_e)^{1/2}$. This is a good approximation for $m_i \gg m_e$, such as ionized matter, but not for situations like electron-hole plasmas in semiconductors, or electron-positron plasmas. The resulting electric field is:

$$\vec{E}_1 = -\frac{\gamma_e T_e}{n_{e0}e}\nabla n_{e1}$$

Since we have already solved for the electric field, we cannot also find it from Poisson's equation. The ion momentum equation now relates n_{i1} for each species to n_{e1}:

$$(-m_i\partial_{tt} + \gamma_i T_i\nabla^2)n_{i1} = -\gamma_e T_e\nabla^2 n_{e1}$$

We arrive at a dispersion relation via Poisson's equation:

$$\frac{\epsilon_0}{e}\nabla\cdot\vec{E}_1 = \left[\sum_{i=1}^{N}n_{i0}Z_i - n_{ne0}\right] + \left[\sum_{i=1}^{N}n_{i1}Z_i - n_{e1}\right]$$

The first bracketed term on the right is zero by assumption (charge-neutral equilibrium). We substitute for the electric field and rearrange to find:

$$(1 - \gamma_e\lambda_{De}^2\nabla^2)n_{e1} = \sum_{i=1}^{N}Z_i n_{i1}.$$

$\lambda_{De}^2 \equiv \epsilon_0 T_e/(n_{e0}e^2)$ defines the electron Debye length. The second term on the left arises from the $\nabla\cdot\vec{E}$ term, and reflects the degree to which the perturbation is not charge-neutral. If $k\lambda_{De}$ is small we may drop this term. This approximation is sometimes called the plasma approximation.

We now work in Fourier space, and write each order-1 field as $X_1 = \tilde{X}_1\exp i(\vec{k}\cdot\vec{x} - \omega t)$ + c.c. We drop the tilde since all equations now apply to the Fourier amplitudes, and find,

$$n_{i1} = \gamma_e T_e Z_i\frac{n_{i0}}{n_{e0}}[m_i v_s^2 - \gamma_i T_i]^{-1}n_{e1}$$

$v_s = \omega/k$ is the wave phase velocity. Substituting this into Poisson's equation gives us an expression where each term is proportional to n_{e1}. To find the dispersion relation for natural modes, we look for solutions for n_{e1} nonzero and find:

$$\gamma_e T_e\left\langle\frac{Z_i^2}{m_i v_s^2 - \gamma_i T_i}\right\rangle = \langle Z_i\rangle(1 + \gamma_e k^2\lambda_{De}^2).$$

$n_{i1} = f_i n_{I1}$ where $n_{I1} = \Sigma_i n_{i1},$, so the ion fractions satisfy $\Sigma_i f_i = 1$ and $\langle X_i \rangle \equiv \Sigma_i f_i X_i$ is the average over ion species. A unitless version of this equation is

$$\frac{\gamma_e}{\langle Z_i \rangle} \left\langle \frac{Z_i^2 / A_i}{u^2 - \tau_i} \right\rangle = 1 + \gamma_e k^2 \lambda_{De}^2$$

with $A_i = m_i / m_u$. m_u, is the atomic mass unit, $u^2 = m_u v_s^2 / T_e$, and

$$\tau_i = \frac{\gamma_i T_i}{A_i T_e}$$

If $k\lambda_{De}$ is small (the plasma approximation), we can neglect the second term on the right-hand side, and the wave is dispersionless $\omega = v_s k$ with v_s independent of k.

Dispersion Relation

The general dispersion relation given above for ion acoustic waves can be put in the form of an order-N polynomial (for N ion species) in u^2. All of the roots should be real-positive, since we have neglected damping. The two signs of u correspond to right- and left-moving waves. For a single ion species,

$$v_s^2 = \frac{\gamma_e Z_i T_e}{m_i} \frac{1}{1 + \gamma_e (k\lambda_{De})^2} + \frac{\gamma_i T_i}{m_i} = \frac{\gamma_e Z_i T_e}{m_i} \left[\frac{1}{1 + \gamma_e (k\lambda_{De})^2} + \frac{\gamma_i T_i}{Z_i \gamma_e T_e} \right]$$

We now consider multiple ion species, for the common case $T_i \ll T_e$. For $T_i = 0$, the dispersion relation has N-1 degenerate roots $u^2 = 0,$, and one non-zero root

$$v_s^2 (T_i = 0) \equiv \frac{\gamma_e T_e / m_u}{1 + \gamma_e (k\lambda_{De})^2} \frac{\langle Z_i^2 / A_i \rangle}{\langle Z_i \rangle}$$

This non-zero root is called the "fast mode", since v_s is typically greater than all the ion thermal speeds. The approximate fast-mode solution for $T_i \ll T_e$ is

$$v_s^2 \approx v_s^2 (T_i = 0) + \frac{\langle Z_i^2 \gamma_i T_i / A_i^2 \rangle}{m_u \langle Z_i^2 / A_i \rangle}$$

The N-1 roots that are zero for $T_i = o$ arc called "slow modes", since v_s can be comparable to or less than the thermal speed of one or more of the ion species.

A case of interest to nuclear fusion is an equimolar mixture of deuterium and tritium ions $(f_D = f_T = 1/2)$. Let us specialize to full ionization $(Z_D = Z_T = 1)$, equal temperatures $(T_e = T_i)$, polytrope exponents $\gamma_e = 1, \gamma_i = 3$, and neglect the $(k\lambda_{De})^2$ contribution. The dispersion relation becomes a quadratic in v_s^2, namely:

$$2 A_D A_T u^4 - 7(A_D + A_T) u^2 + 24 = 0$$

Using $(A_D, A_T) = (2.01, 3.02)$ we find the two roots are $u^2 = (1.10, 1.81)$.

Another case of interest is one with two ion species of very different masses. An example is a mixture of gold (A=197) and boron (A=10.8), which is currently of interest in hohlraums for laser-driven inertial fusion research. For a concrete example, consider $\gamma_e = 1$ and $\gamma_i = 3, T_i = T_e / 2$ for both ion species, and charge states Z=5 for boron and Z=50 for gold. We leave the boron atomic fraction f_B unspecified (note $f_{Au} = 1 - f_B$). Thus, $\bar{Z} = 50 - 45 f_B, \tau_B = 0.139, \tau_{Au} = 0.00761, F_B = 2.31 f_B / \bar{Z}$, and $F_{Au} = 12.69(1 - f_B) / \bar{Z}$.

Damping

Ion acoustic waves are damped both by Coulomb collisions and collisionless Landau damping. The Landau damping occurs on both electrons and ions, with the relative importance depending on parameters.

ALFVÉN WAVE

In plasma physics, an Alfvén wave, named after Hannes Alfvén, is a type of magneto-hydrodynamicwave in which ions oscillate in response to a restoring force provided by an effective tension on the magnetic field lines.

An Alfvén wave in a plasma is a low-frequency (compared to the ion cyclotron frequency) travelling oscillation of the ions and the magnetic field. The ion mass density provides the inertia and the magnetic field line tension provides the restoring force.

The wave propagates in the direction of the magnetic field, although waves exist at oblique incidence and smoothly change into the magnetosonic wave when the propagation is perpendicular to the magnetic field.

The motion of the ions and the perturbation of the magnetic field are in the same direction and transverse to the direction of propagation. The wave is dispersionless.

Alfvén Velocity

The low-frequency relative permittivity ϵ of a magnetized plasma is given by,

$$\epsilon = 1 + \frac{1}{B^2} c^2 \mu_0 \rho$$

Where B is the magnetic field strength, c is the speed of light, μ_0 is the permeability of the vacuum, and $\rho = \Sigma n_s m_s$ is the total mass density of the charged plasma particles. Here, s goes over all plasma species, both electrons and (few types of) ions.

Therefore, the phase velocity of an electromagnetic wave in such a medium is:

$$v = \frac{c}{\sqrt{\epsilon}} = \frac{c}{\sqrt{1 + \frac{1}{B^2} c^2 \mu_0 \rho}}$$

or

$$v = \frac{v_A}{\sqrt{1 + \frac{1}{c^2} v_A^2}}$$

where:

$$v_A = \frac{B}{\sqrt{\mu_0 \rho}},$$

is the Alfvén velocity. If $v_A \ll c$, then $v \approx v_A$. On the other hand, when $v_A \to c$, then $v \approx c$. That is, at high field or low density, the velocity of the Alfvén wave approaches the speed of light, and the Alfvén wave becomes an ordinary electromagnetic wave.

Neglecting the contribution of the electrons to the mass density and assuming that there is a single ion species, we get

$$v_A = \frac{B}{\sqrt{\mu_0 n_i m_i}} \quad in\ SI$$

$$v_A = \frac{B}{\sqrt{4\pi n_i m_i}} \quad in\ Gauss$$

$$v_A \approx (2.18 \times 10^{11}\ \mathrm{cm/s})(m_i / m_p)^{-1/2} (n_i / \mathrm{cm}^{-3})^{-1/2} (B / \mathrm{gauss})$$

where n_i is the ion number density and m_i is the ion mass.

Alfvén Time

In plasma physics, the Alfvén time τ_A is an important timescale for wave phenomena. It is related to the Alfvén velocity by:

$$\tau_A = \frac{a}{v_A}$$

where n_i denotes the characteristic scale of the system. For example, a could be the minor radius of the torus in a tokamak.

Relativistic Case

In 1993, Gedalin derived the Alfvén wave velocity using relativistic magnetohydrodynamicsto be,

$$v = \frac{c}{\sqrt{1 + \dfrac{e + P}{2P_m}}}$$

where is the total energy density of plasma particles, P is the total plasma pressure, and $P_m = \dfrac{B^2}{2\mu_0}$ is the magnetic pressure. In the non-relativistic limit $P \ll e \approx \rho c^2$,

References

- Kaghashvili, Edisher Kh.; Quinn, Richard A.; Hollweg, Joseph V. (2009). "Driven Waves as a Diagnostics Tool in the Solar Corona". The Astrophysical Journal. 703 (2): 1318. Bibcode:2009ApJ...703.1318K. doi:10.1088/0004-637x/703/2/1318

- Andreev, A. A. (2000), An Introduction to Hot Laser Plasma Physics, Huntington, New York: Nova Science Publishers, Inc., ISBN 978-1-56072-803-0

- McIntosh; et al. (2011). "Alfvenic waves with sufficient energy to power the quiet solar corona and fast solar wind". Nature. 475 (7357): 477–480. Bibcode:2011Natur.475.477M. doi:10.1038/nature10235. PMID 21796206

- Karen Fox (27 July 2011). "SDO Spots Extra Energy in the Sun's Corona". NASA. Retrieved 2 April 2012.

- Doorsselaere, T. Van; Nakariakov, V. M.; Verwichte, E. (2008). "Detection of Waves in the Solar Corona: Kink or Alfvén?". The Astrophysical Journal Letters. 676 (1): L73

Plasma Diagnostics

Plasma diagnostics refers to the methods, techniques and instruments that are used for the measurement of varied properties of plasma. Some of the most common invasive probe methods used are Langmuir probe, ball-pen probe and Faraday cup. The chapter closely examines these key concepts of plasma diagnostics to provide an extensive understanding of the subject.

Characterization of plasma sources is an important task for process understanding, analysis, and control. The characterization of the plasma sources involves the use of many sophisticated diagnostic tools.

The techniques developed for diagnosing the properties of plasma is known as plasma diagnostics.

The main objective of the plasma diagnostics is to deduce informations about the state of the plasma from practical observations of physical processes and their effects. Accurate and reliable measurements of the plasma condition in the various plasma experiments are an important objective for making a significant progress in the field. Since the plasma behavior is dependent on the condition in which it is produced, so it demands knowledge of many plasma quantities for its description as well as for its comparison with relevant theory. Particularly in fusion experiments many important quantities are either measured with poor accuracy or even not measured at all. Plasma research and particularly fusion research cover a wide area of basic physics and use most advanced technology.

The plasma diagnostics can be cetegorized in various ways. First of all one can consider cold and hot plasma. Material probes such as the most common Langmuir probe can be inserted in the cold plasma. However in the hot plasma ($T_e \geq$ few keV) the use of material probes is limited to the extreme edge of the plasma ($T_e \leq 50eV$) and the non − invasive techniques must be used to diagnose other plasma regions. These includes passive methods, which detects radiation or particles emitted spontaneously by the plasma, where as in active methods radiations or particles produced by external sources are used to probe the plasma. Another way of grouping them are on the basis of measured plasma parameters such as plasma density and temperature etc.

INVASIVE PROBE METHODS

Langmuir Probe

A Langmuir probe is a device used to determine the electron temperature, electron density, and electric potential of a plasma. It works by inserting one or more electrodes into a plasma, with a constant or time-varying electric potential between the various electrodes or between them and the surrounding vessel. The measured currents and potentials in this system allow the determination of the physical properties of the plasma.

I-V Characteristic of the Debye Sheath

The beginning of Langmuir probe theory is the *I-V* characteristic of the Debye sheath, that is, the current density flowing to a surface in a plasma as a function of the voltage drop across the sheath. The analysis presented here indicates how the electron temperature, electron density, and plasma potential can be derived from the *I-V* characteristic. In some situations a more detailed analysis can yield information on the ion density (T_i), the ion temperature $f_e(v)$, or the electron energy distribution function (EEDF).

Ion Saturation Current Density

Consider first a surface biased to a large negative voltage. If the voltage is large enough, essentially all electrons (and any negative ions) will be repelled. The ion velocity will satisfy the Bohm sheath criterion, which is, strictly speaking, an inequality, but which is usually marginally fulfilled. The Bohm criterion in its marginal form says that the ion velocity at the sheath edge is simply the sound speed given by:

$$c_s = \sqrt{k_B(ZT_e + \gamma_i T_i)/m_i}.$$

The ion temperature term is often neglected, which is justified if the ions are cold. Even if the ions are known to be warm, the ion temperature is usually not known, so it is usually assumed to be simply equal to the electron temperature. In that case, consideration of finite ion temperature only results in a small numerical factor. Z is the (average) charge state of the ions, and γ_i is the adiabatic coefficient for the ions. The proper choice of γ_i is a matter of some contention. Most analyses use $\gamma_i = 1$, corresponding to isothermal ions, but some kinetic theory suggests that $\gamma_i = 3$, corresponding to one degree of freedom is more appropriate. For $Z = 1$ and $T_i = T_e$ using the larger value results in the conclusion that the density is $\sqrt{2}$ times smaller. Uncertainties of this magnitude arise several places in the analysis of Langmuir probe data and are very difficult to resolve.

The charge density of the ions depends on the charge state Z, but quasineutrality allows one to write it simply in terms of the electron density as en_e.

Using these results we have the current density to the surface due to the ions. The current density at large negative voltages is due solely to the ions and, except for possible sheath expansion effects, does not depend on the bias voltage, so it is referred to as the ion saturation current density and is given by:

$j_i^{max} = q_e n_e c_s$ where q_e is the charge of an electron, n_e is the number density of electrons, and c_s is as defined above.

The plasma parameters, in particular, the density, are those at the sheath edge.

Exponential Electron Current

As the voltage of the Debye sheath is reduced, the more energetic electrons are able to overcome the potential barrier of the electrostatic sheath. We can model the electrons at the sheath edge with a Maxwell–Boltzmann distribution, i.e.,

$$f(v_x)dv_x \propto e^{-\frac{1}{2}m_e v_x^2 / k_B T_e},$$

except that the high energy tail moving away from the surface is missing, because only the lower energy electrons moving toward the surface are reflected. The higher energy electrons overcome the sheath potential and are absorbed. The mean velocity of the electrons which are able to overcome the voltage of the sheath is,

$$\langle v_e \rangle = \frac{\int_{v_{e0}}^{\infty} f(v_x)v_x dv_x}{\int_{-\infty}^{\infty} f(v_x)dv_x},$$

where the cut-off velocity for the upper integral is,

$$v_{e0} = \sqrt{2q_e \Delta V / m_e}.$$

ΔV is the voltage across the Debye sheath, that is, the potential at the sheath edge minus the potential of the surface. For a large voltage compared to the electron temperature, the result is,

$$\langle v_e \rangle = \sqrt{\frac{k_B T_e}{2\pi m_e}} e^{-q_e \Delta V / k_B T_e}.$$

With this expression, we can write the electron contribution to the current to the probe in terms of the ion saturation current as:

$$j_e = j_i^{max} \sqrt{m_i / 2\pi m_e} \, e^{-q_e \Delta V / k_B T_e},$$

valid as long as the electron current is not more than two or three times the ion current.

Floating Potential

The total current, of course, is the sum of the ion and electron currents:

$$j = j_i^{max}\left(-1 + \sqrt{m_i/2\pi m_e}\, e^{-q_e \Delta V/k_B T_e}\right).$$

We are using the convention that current *from* the surface into the plasma is positive. An interesting and practical question is the potential of a surface to which no net current flows. It is easily seen from the above equation that,

$$\Delta V = (k_B T_e/e)(1/2)\ln(m_i/2\pi m_e).$$

If we introduce the ion reduced mass $\mu_i = m_i/m_e$, we can write:

$$\Delta V = (k_B T_e/e)(2.8 + 0.5\ln \mu_i)$$

Since the floating potential is the experimentally accessible quantity, the current (below electron saturation) is usually written as:

$$j = j_i^{max}\left(-1 + e^{q_e(V_0 - \Delta V)/k_B T_e}\right).$$

Electron Saturation Current

When the electrode potential is equal to or greater than the plasma potential, then there is no longer a sheath to reflect electrons, and the electron current saturates. Using the Boltzmann expression for the mean electron velocity given above with $v_{e0} = 0$ and setting the ion current to zero, the electron saturation current density would be,

$$j_e^{max} = j_i^{max}\sqrt{m_i/\pi m_e} = j_i^{max}\left(24.2\sqrt{\mu_i}\right)$$

Although this is the expression usually given in theoretical discussions of Langmuir probes, the derivation is not rigorous and the experimental basis is weak. The theory of double layers typically employs an expression analogous to the Bohm criterion, but with the roles of electrons and ions reversed, namely,

$$j^{max} = q\, n\, \sqrt{k\,(\gamma T + T)/m} = j^{max}\sqrt{m/m} = j^{max}\left(42.8\sqrt{\mu}\right)$$

where the numerical value was found by taking $T_i = T_e$ and $\gamma_i = \gamma_e$.

In practice, it is often difficult and usually considered uninformative to measure the electron saturation current experimentally. When it is measured, it is found to be highly variable and generally much lower (a factor of three or more) than the value given above. Often a clear saturation is not seen at all. Understanding electron saturation is one of the most important outstanding problems of Langmuir probe theory.

Effects of the Bulk Plasma

The Debye sheath theory explains the basic behavior of Langmuir probes, but is not complete. Merely inserting an object like a probe into a plasma changes the density, temperature, and potential at the sheath edge and perhaps everywhere. Changing the voltage on the probe will also, in general, change various plasma parameters. Such effects are less well understood than sheath physics, but they can at least in some cases be roughly accounted.

Pre-sheath

The Bohm criterion requires the ions to enter the Debye sheath at the sound speed. The potential drop that accelerates them to this speed is called the pre-sheath. It has a spatial scale that depends on the physics of the ion source but which is large compared to the Debye length and often of the order of the plasma dimensions. The magnitude of the potential drop is equal to (at least),

$$\Phi_{pre} = \frac{\frac{1}{2} m_i c_s^2}{Ze} = k_B (T_e + Z\gamma_i T_i)/(2Ze)$$

The acceleration of the ions also entails a decrease in the density, usually by a factor of about 2 depending on the details.

Resistivity

Collisions between ions and electrons will also affect the I-V characteristic of a Langmuir probe. When an electrode is biased to any voltage other than the floating potential, the current it draws must pass through the plasma, which has a finite resistivity. The resistivity and current path can be calculated with relative ease in an unmagnetized plasma. In a magnetized plasma, the problem is much more difficult. In either case, the effect is to add a voltage drop proportional to the current drawn, which shears the characteristic. The deviation from an exponential function is usually not possible to observe directly, so that the flattening of the characteristic is usually misinterpreted as a larger plasma temperature. Looking at it from the other side, any measured I-V characteristic can be interpreted as a hot plasma, where most of the voltage is dropped in the Debye sheath, or as a cold plasma, where most of the voltage is dropped in the bulk plasma. Without quantitative modeling of the bulk resistivity, Langmuir probes can only give an upper limit on the electron temperature.

Sheath Expansion

It is not enough to know the current *density* as a function of bias voltage since it is the *absolute* current which is measured. In an unmagnetized plasma, the current-collecting area is usually taken to be the exposed surface area of the electrode. In a magnetized plasma, the projected area is taken, that is, the area of the electrode as viewed along the magnetic field. If the electrode is not shadowed by a wall or other nearby object, then the area must be doubled to account for current coming along the field from both sides.

If the electrode dimensions are not small in comparison to the Debye length, then the size of the electrode is effectively increased in all directions by the sheath thickness. In a magnetized plasma, the electrode is sometimes assumed to be increased in a similar way by the ion Larmor radius.

The finite Larmor radius allows some ions to reach the electrode that would have otherwise gone past it. The details of the effect have not been calculated in a fully self-consistent way.

If we refer to the probe area including these effects as A_{eff} (which may be a function of the bias voltage) and make the assumptions:

$$T_i \quad T_e,$$

$$Z = 1$$

$$\gamma_i = 3, \; and$$

$$+ n_{e,sh} = 0.5 n_e,$$

and ignore the effects of bulk resistivity, and electron saturation, then the *I-V* characteristic becomes,

$$I = I_i^{max}(-1 + e^{q_e(V_{pr}-V_{fl})/(k_B T_e)}),$$

where,

$$I_i^{max} = q_e n_e \sqrt{k_B T_e / m_i} \, A_{eff}.$$

Magnetized Plasmas

The theory of Langmuir probes is much more complex when the plasma is magnetized. The simplest extension of the unmagnetized case is simply to use the projected area rather than the surface area of the electrode.

For a long cylinder far from other surfaces, this reduces the effective area by a factor of $\pi/2 = 1.57$. As mentioned before, it might be necessary to increase the radius by about the thermal ion Larmor radius, but not above the effective area for the unmagnetized case.

The use of the projected area seems to be closely tied with the existence of a magnetic sheath. Its scale is the ion Larmor radius at the sound speed, which is normally between the scales of the Debye sheath and the pre-sheath. The Bohm criterion for ions entering the magnetic sheath applies to the motion along the field, while at the entrance to the Debye sheath it applies to the motion normal to the surface. This results in a reduction of the density by the sine of the angle between the field and the surface. The associated increase in the Debye length must be taken into account when considering ion non-saturation due to sheath effects.

Especially interesting and difficult to understand is the role of cross-field currents. Naively, one would expect the current to be parallel to the magnetic field along a flux tube. In many geometries, this flux tube will end at a surface in a distant part of the device, and this spot should itself exhibit an *I-V* characteristic. The net result would be the measurement of a double-probe characteristic; in other words, electron saturation current equal to the ion saturation current.

When this picture is considered in detail, it is seen that the flux tube must charge up and the surrounding plasma must spin around it. The current into or out of the flux tube must be associated with a force that slows down this spinning. Candidate forces are viscosity, friction with neutrals, and inertial forces associated with plasma flows, either steady or fluctuating. It is not known which force is strongest in practice, and in fact it is generally difficult to find any force that is powerful enough to explain the characteristics actually measured.

It is also likely that the magnetic field plays a decisive role in determining the level of electron saturation, but no quantitative theory is as yet available.

Electrode Configurations

Once one has a theory of the *I-V* characteristic of an electrode, one can proceed to measure it and then fit the data with the theoretical curve to extract the plasma parameters. The straightforward way to do this is to sweep the voltage on a single electrode, but, for a number of reasons, configurations using multiple electrodes or exploring only a part of the characteristic are used in practice.

Single Probe

The most straightforward way to measure the *I-V* characteristic of a plasma is with a single probe, consisting of one electrode biased with a voltage ramp relative to the vessel. The advantages are simplicity of the electrode and redundancy of information, i.e. one can check whether the *I-V* characteristic has the expected form. Potentially additional information can be extracted from details of the characteristic. The disadvantages are more complex biasing and measurement electronics and a poor time resolution. If fluctuations are present (as they always are) and the sweep is slower than the fluctuation frequency (as it usually is), then the *I-V* is the *average* current as a function of voltage, which may result in systematic errors if it is analyzed as though it were an instantaneous *I-V*. The ideal situation is to sweep the voltage at a frequency above the fluctuation frequency but still below the ion cyclotron frequency. This, however, requires sophisticated electronics and a great deal of care.

Double Probe

An electrode can be biased relative to a second electrode, rather than to the ground. The theory is similar to that of a single probe, except that the current is limited to the ion saturation current for both positive and negative voltages. In particular, if V_{bias} is the voltage applied between two identical electrodes, the current is given by;

$$I = I_i^{max}\left(-1+e^{q_e(V_2-V_{fl})/k_BT_e}\right)=-I_i^{max}\left(-1+e^{q_e(V_1-V_{fl})/k_BT_e}\right),$$

which can be rewritten using $V_{bias}=V_2-V_1$ as a hyperbolic tangent:

$$I = I_i^{max}\tanh\left(\frac{1}{2}\frac{q_eV_{bias}}{k_BT_e}\right).$$

One advantage of the double probe is that neither electrode is ever very far above floating, so the theoretical uncertainties at large electron currents are avoided. If it is desired to sample more of the exponential electron portion of the characteristic, an asymmetric double probe may be used, with one electrode larger than the other. If the ratio of the collection areas is larger than the square root of the ion to electron mass ratio, then this arrangement is equivalent to the single tip probe. If the ratio of collection areas is not that big, then the characteristic will be in-between the symmetric double tip configuration and the single-tip configuration. If A_1 is the area of the larger tip then:

$$I = A_1 J_i^{max}\left[\coth\left(\frac{q_eV_{bias}}{2k_BT_e}\right)+\frac{\left(\dfrac{A_1}{A_2}-1\right)e^{-q_eV_{bias}/2k_BT_e}}{2\sinh\left(\dfrac{q_eV_{bias}}{2k_BT_e}\right)}\right]^{-1}$$

Another advantage is that there is no reference to the vessel, so it is to some extent immune to the disturbances in a radio frequency plasma. On the other hand, it shares the limitations of a single probe concerning complicated electronics and poor time resolution. In addition, the second electrode not only complicates the system, but it makes it susceptible to disturbance by gradients in the plasma.

Triple Probe

An elegant electrode configuration is the triple probe, consisting of two electrodes biased with a fixed voltage and a third which is floating. The bias voltage is chosen to be a few times the electron temperature so that the negative electrode draws the ion saturation current, which, like the floating potential, is directly measured. A common rule of thumb for this voltage bias is 3/e times the expected electron temperature. Because the biased tip configuration is floating, the positive probe can draw at most an electron current only equal in magnitude and opposite in polarity to the ion saturation current drawn by the negative probe, given by:

$$-I_+ = I_- = I_i^{max}$$

and as before the floating tip draws effectively no current:

$$I_{fl} = 0.$$

Assuming that: 1.) The electron energy distribution in the plasma is Maxwellian, 2.) The mean free path of the electrons is greater than the ion sheath about the tips and larger than the probe radius, and 3.) the probe sheath sizes are much smaller than the probe separation, then the current to any probe can be considered composed of two parts – the high energy tail of the Maxwellian electron distribution, and the ion saturation current:

$$I_{probe} = -I_e e^{-q_e V_{probe}/(kT_e)} + I_i^{max}$$

where the current I_e is thermal current. Specifically,

$$I_e = SJ_e = Sn_e q_e \sqrt{kT_e / 2\pi m_e}$$

where S is surface area, J_e is electron current density, and n_e is electron density.

Assuming that the ion and electron saturation current is the same for each probe, then the formulas for current to each of the probe tips take the form,

$$I_+ = -I_e e^{-q_e V_+/(kT_e)} + I_i^{max}$$
$$I_- = -I_e e^{-q_e V_-/(kT_e)} + I_i^{max}$$
$$I_{fl} = -I_e e^{-q_e V_{fl}/(kT_e)} + I_i^{max}.$$

It is then simple to show,

$$\left(I_+ - I_{fl}\right)/\left(I_+ - I_-\right) = \left(1 - e^{-q_e(V_{fl}-V_+)/(kT_e)}\right)/\left(1 - e^{-q_e(V_- - V_+)/(kT_e)}\right)$$

but the relations from above specifying that I_+=-I_- and I_{fl}=0 give:

$$1/2 = \left(1 - e^{-q_e(V_{fl}-V_+)/(kT_e)}\right)/\left(1 - e^{-q_e(V_- - V_+)/(kT_e)}\right),$$

a transcendental equation in terms of applied and measured voltages and the unknown T_e that in the limit $q_e V_{Bias} = q_e(V_+ - V_-) \gg$ becomes,

$$(V_+ - V_{fl}) = (k_B T_e / q_e) \ln 2.$$

That is, the voltage difference between the positive and floating electrodes is proportional to the electron temperature. (This was especially important in the sixties and seventies before sophisticated data processing became widely available).

More sophisticated analysis of triple probe data can take into account such factors as incomplete saturation, non-saturation, unequal areas.

Triple probes have the advantage of simple biasing electronics (no sweeping required), simple data analysis, excellent time resolution, and insensitivity to potential fluctuations (whether imposed by an rf source or inherent fluctuations). Like double probes, they are sensitive to gradients in plasma parameters.

Special Arrangements

Arrangements with four (tetra probe) or five (penta probe) have sometimes been used, but the advantage over triple probes has never been entirely convincing. The spacing between probes must be larger than the Debye length of the plasma to prevent an overlapping Debye sheath.

A pin-plate probe consists of a small electrode directly in front of a large electrode, the idea being that the voltage sweep of the large probe can perturb the plasma potential at the sheath edge and thereby aggravate the difficulty of interpreting the *I-V* characteristic. The floating potential of the small electrode can be used to correct for changes in potential at the sheath edge of the large probe. Experimental results from this arrangement look promising, but experimental complexity and residual difficulties in the interpretation have prevented this configuration from becoming standard.

Various geometries have been proposed for use as ion temperature probes, for example, two cylindrical tips that rotate past each other in a magnetized plasma. Since shadowing effects depend on the ion Larmor radius, the results can be interpreted in terms of ion temperature. The ion temperature is an important quantity that is very difficult to measure. Unfortunately, it is also very difficult to analyze such probes in a fully self-consistent way.

Emissive probes use an electrode heated either electrically or by the exposure to the plasma. When the electrode is biased more positive than the plasma potential, the emitted electrons are pulled back to the surface so the *I-V* characteristic is hardly changed. As soon as the electrode is biased negative with respect to the plasma potential, the emitted electrons are repelled and contribute a large negative current.

The onset of this current or, more sensitively, the onset of a discrepancy between the characteristics of an unheated and a heated electrode, is a sensitive indicator of the plasma potential.

To measure fluctuations in plasma parameters, arrays of electrodes are used, usually one – but occasionally two-dimensional. A typical array has a spacing of 1 mm and a total of 16 or 32 electrodes. A simpler arrangement to measure fluctuations is a negatively biased electrode flanked by two floating electrodes. The ion-saturation current is taken as a surrogate for the density and the floating potential as a surrogate for the plasma potential. This allows a rough measurement of the turbulent particle flux,

$$\Phi_{turb} = \langle \tilde{n}_e \tilde{v}_{E\times B} \rangle \propto \langle \tilde{I}_i^{max} (\tilde{V}_{fl,2} - \tilde{V}_{fl,1}) \rangle$$

Cylindrical Langmuir Probe in Electron Flow

Most often, the Langmuir probe is a small size electrode inserted in plasma and connected through an external (with respect to plasma) electric circuit with the electrode of a large surface area contacting with the same plasma (very often it is metallic wall of a chamber containing plasma) to obtain the probe I-V characteristic $i(V)$. The characteristic $i(V)$ is measured by sweeping the voltage V of scanning generator (inserted in the probe circuit) with simultaneous measuring of the probe current.

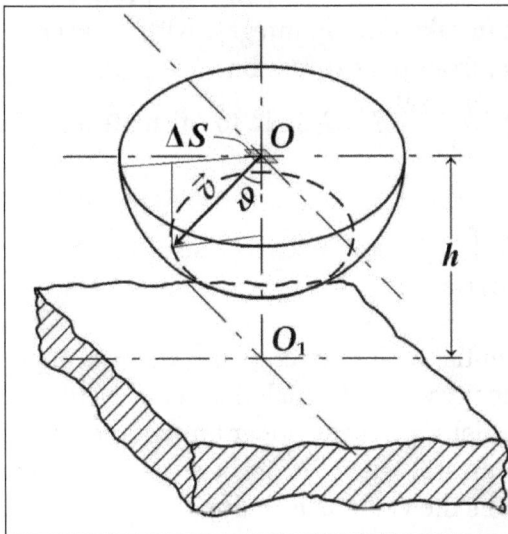

Illustration to Langmuir Probe I-V Characteristic Derivation.

Relations between the probe I-V characteristic and parameters of isotropic plasma were found by the Irving Langmuir and they can be derived most elementary for the planar probe of a large surface area S_z (ignoring the edge effects problem). Let us choose the point O in plasma at the distance h from the probe surface where electric field of the probe is negligible while each electron of plasma passing this point could reach the probe surface without collisions with plasma components: $\lambda_D \ll \lambda_{Te}.\lambda_D$, is the Debye length and λ_{Te} is the electron free path calculated for its total cross section with plasma components. In the vicinity of the point O we can imagine a small element of the surface area ΔS parallel to the probe surface. The elementary current di of plasma electrons passing throughout ΔS in a direction of the probe surface can be written in the form,

$$di = q_e \Delta S dn(v, \vartheta) v \cos \vartheta,$$

where v is a scalar of the electron thermal velocity vector $\vec{v}$,

$$dn(v, \vartheta) = n f(v) \frac{2\pi \sin \vartheta}{4\pi} dv d\vartheta,$$

$2\pi \sin \vartheta d\vartheta$ is the element of the solid angle with its relative value $2\pi \sin \vartheta d\vartheta / 4\pi.\vartheta$ is the angle between perpendicular to the probe surface recalled from the point O and the radius-vector of the electron thermal velocity $\vec{v}$ forming a spherical layer of thickness dv in velocity space, and $f(v)$ is the electron distribution function normalized to unity

$$\int_0^\infty f(v) dv = 1.$$

Taking into account uniform conditions along the probe surface (boundaries are excluded), $\Delta S \to S_z$, we can take double integral with respect to the angle ϑ and with respect to the velocity v, from the expression $di = q_e \Delta S dn(v, \vartheta) v \cos \vartheta$, after substitution eq. $dn(v, \vartheta) = n f(v) \frac{2\pi \sin \vartheta}{4\pi} dv d\vartheta$, in it, to calculate a total electron current on the probe

$$i(v) = q_e n S_z \frac{1}{4\pi} \int_{\sqrt{2q_e V/m}}^\infty f(v) dv \int_0^\zeta v \cos \vartheta 2\pi \sin \vartheta d\vartheta.$$

where V is the probe potential with respect to the potential of plasma $V = 0, \sqrt{2q_e V/m}$, is the lowest electron velocity value at which the electron still could reach the probe surface charged to the potential $V.\zeta$, is the upper limit of the angle ϑ at which the electron having initial velocity v can still reach the probe surface with a zero-value of its velocity at this surface. That means the value ζ is defined by the condition,

$$v \cos \zeta = \sqrt{2q_e V/m}.$$

Deriving the value ζ from eq. above and substituting it in eq.,

$$i(v) = q_e n S_z \frac{1}{4\pi} \int_{\sqrt{2q_e V/m}}^\infty f(v) dv \int_0^\zeta v \cos \vartheta 2\pi \sin \vartheta d\vartheta$$

we can obtain the probe I-V characteristic (neglecting the ion current) in the range of the probe potential $-\infty < V \leq 0$ in the form,

$$i(V) = \frac{q_e n S_z}{4} \int\limits_{\sqrt{2q_e V/m}}^{\infty} f(v)\left(1 - \frac{2q_e V}{mv^2}\right) v dv.$$

Differentiating eq. above twice with respect to the potential V, one can find the expression describing the second derivative of the probe I-V characteristic (obtained firstly by M. J. Druyvestein,

$$i''(V) = \frac{q_e^2 n S_z}{4m} \frac{1}{V} f\left(\sqrt{2q_e V/m}\right)$$

defining the electron distribution function over velocity $f\left(\sqrt{2q_e V/m}\right)$ in the evident form. M. J. Druyvestein has shown in particular that eqs.,

$$i(V) = \frac{q_e n S_z}{4} \int\limits_{\sqrt{2q_e V/m}}^{\infty} f(v)\left(1 - \frac{2q_e V}{mv^2}\right) v dv.$$

and $i''(V) = \frac{q_e^2 n S_z}{4m} \frac{1}{V} f\left(\sqrt{2q_e V/m}\right)$ are valid for description of operation of the probe of any arbitrary convex geometrical shape. Substituting the Maxwellian distribution function:

$$f^{(0)}(v) = \frac{4}{\sqrt{\pi}} \frac{v^2}{v_p^3} \exp\left(-v^2/v_p^2\right),$$

where $v_p = \langle v \rangle \sqrt{\pi}/2$ is the most probable velocity, in eq.,

$$i(V) = \frac{q_e n S_z}{4} \int\limits_{\sqrt{2q_e V/m}}^{\infty} f(v)\left(1 - \frac{2q_e V}{mv^2}\right) v dv.$$

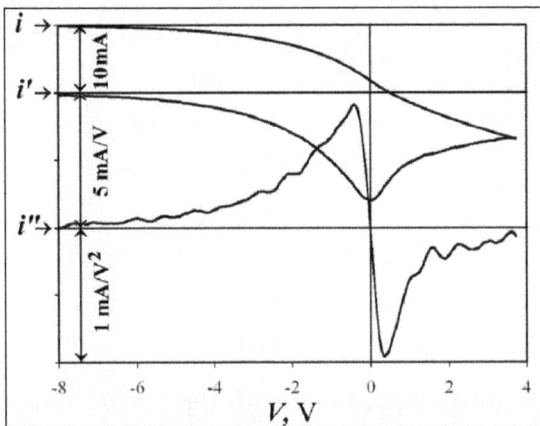

I-V Characteristic of Langmuir Probe in Isotropic Plasma.

we obtain the expression,

$$i^{(0)}(V) = \frac{q_e n \langle v \rangle}{4} S_z \exp\left(-q_e V / \mathcal{E}_p\right)$$

From which the very useful in practice relation follows,

$$\ln\left(i^{(0)}(V) / i^{(0)}(0)\right) = -q_e V / \mathcal{E}_p.$$

allowing one to derive the electron energy $\mathcal{E}_p = k_B T$ (for Maxwellian distribution function only!) by a slope of the probe I-V characteristic in a semilogarithmic scale. Thus in plasmas with isotropic electron distributions, the electron current on $i_{th}(0)$ a surface $S_z = 2\pi r_z l_z$ of the cylindrical Langmuir probe at plasma potential $V = o$ is defined by the average electron thermal velocity $\langle v \rangle$ and can be written down as equation,

$$i_{th}(0) = q_e n \langle v \rangle \frac{1}{4} \times 2\pi r_z l_z,$$

where n is the electron concentration, r_z is the probe radius, and l_z is its length. It is *obvious that if plasma electrons form an electron wind (flow) across the cylindrical* probe axis with a velocity $v_d \gg \langle v \rangle$, the expression,

$$i_d = en v_d \times 2 r_z l_z$$

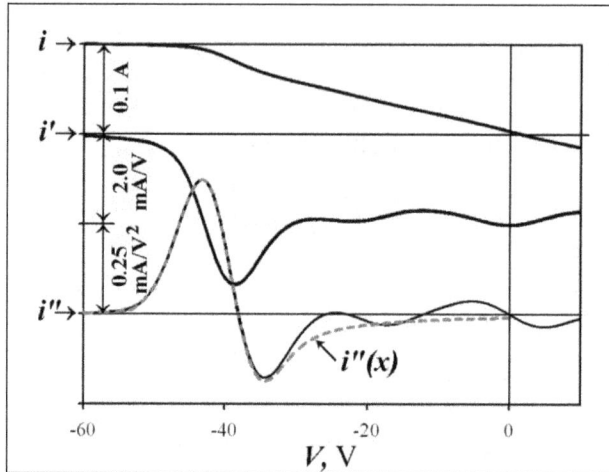

I-V Characteristic of the cylindrical
probe in crossing electron wind.

holds true. In plasmas produced by gas-discharge arc sources as well as inductively coupled sources, the electron wind can develop the Mach number $M^{(0)} = v_d / \langle v \rangle = (\sqrt{\pi} / 2)\alpha \gtrsim 1$. Here the parameter α is introduced along with the Mach number for simplification of mathematical expressions. Note that $(\sqrt{\pi} / 2)\langle v \rangle = v_p$, where p is the most probable velocity for the Maxwellian distribution function, so that $\alpha = v_d / v_p$. Thus the general case where $\alpha \gtrsim 1$ is of the theoretical and practical interest. Corresponding physical and mathematical considerations presented has shown that at the Maxwellian distribution function of the electrons in a reference system moving with the velocity v_d *across axis of the cylindrical* probe set at plasma potential $V= 0$. the electron current on the probe can be written down in the form,

$$\frac{i(0)}{enS_z} = \frac{\langle v \rangle}{4} \exp(-\alpha^2 / 2) I_0(\alpha^2 / 2) \left(1 + \alpha^2 \left(1 + I_1(\alpha^2 / 2) / I_0(\alpha^2 / 2)\right)\right),$$

where I_0 and I_1 are Bessel functions of imaginary arguments and eq. above is reduced to eq. $i_{th}(0) = q_e n \langle v \rangle \frac{1}{4} \times 2\pi r_z l_z$, at $\alpha \to 0$ being reduced to eq. $i_d = env_d \times 2r_z l_z$ at $\alpha \to \infty$. The second derivative of the probe I-V characteristic $i''(V)$ with respect to the probe potential V can be presented in this case in the form,

$$i''(x) = enS_z \frac{v_p}{2\pi^{3/2}(\mathcal{E}_p / e)^2} \frac{1}{\sqrt{x}} \int_0^\pi (\sqrt{x} - \cos\varphi) \exp\left(-\alpha^2(\sqrt{x} - \cos\varphi)\right) d\varphi,$$

where:

$$x = \frac{1}{\alpha^2} \frac{V}{\mathcal{E}_p / e}$$

and the electron energy $\mathcal{E}_p / e$ is expressed in eV.

All parameters of the electron populationn $n. \alpha \langle v \rangle$ and v_p in plasma can be derived from the experimental probe I-V characteristic second derivative $i''(V)$ by its least square best fitting with the theoretical curve expressed by.

Practical Considerations

For laboratory and technical plasmas, the electrodes are most commonly tungsten or tantalum wires several thousandths of an inch thick, because they have a high melting point but can be made small enough not to perturb the plasma. Although the melting point is somewhat lower, molybdenum is sometimes used because it is easier to machine and solder than tungsten. For fusion plasmas, graphite electrodes with dimensions from 1 to 10 mm are usually used because they can withstand the highest power loads (also sublimating at high temperatures rather than melting), and result in reduced bremsstrahlung radiation (with respect to metals) due to the low atomic number of carbon. The electrode surface exposed to the plasma must be defined, e.g. by insulating all but the tip of a wire electrode. If there can be significant deposition of conducting materials (metals or graphite), then the insulator should be separated from the electrode by a meander to prevent short-circuiting.

In a magnetized plasma, it appears to be best to choose a probe size a few times larger than the ion Larmor radius. A point of contention is whether it is better to use proud probes, where the angle between the magnetic field and the surface is at least 15°, or flush-mounted probes, which are embedded in the plasma-facing components and generally have an angle of 1 to 5 °. Many plasma physicists feel more comfortable with proud probes, which have a longer tradition and possibly are less perturbed by electron saturation effects, although this is disputed. Flush-mounted probes, on the other hand, being part of the wall, are less perturbative. Knowledge of the field angle is necessary with proud probes to determine the fluxes to the wall, whereas it is necessary with flush-mounted probes to determine the density.

In very hot and dense plasmas, as found in fusion research, it is often necessary to limit the thermal load to the probe by limiting the exposure time. A reciprocating probe is mounted on an arm that is moved into and back out of the plasma, usually in about one second by means of either a pneumatic drive or an electromagnetic drive using the ambient magnetic field. Pop-up probes are similar, but the electrodes rest behind a shield and are only moved the few millimeters necessary to bring them into the plasma near the wall.

A Langmuir probe can be purchased off the shelf for on the order of 15,000 U.S. dollars, or they can be built by an experienced researcher or technician. When working at frequencies under 100 MHz, it is advisable to use blocking filters, and take necessary grounding precautions.

In low temperature plasmas, in which the probe does not get hot, surface contamination may become an issue. This effect can cause hysteresis in the I-V curve and may limit the current collected by the probe. A heating mechanism or a glow discharge plasma may be used to clean the probe and prevent misleading results.

Ball-pen Probe

A ball-pen probe is a modified Langmuir probe used to measure the plasma potential in magnetized plasmas. The ball-pen probe balances the electron and ion saturation currents, so that its floating potential is equal to the plasma potential. Because electrons have a much smaller gyroradius than ions, a moving ceramic shield can be used to screen off an adjustable part of the electron current from the probe collector.

Ball-pen probes are used in plasma physics, notably in tokamaks such as CASTOR, (Czech Academy of Sciences Torus) ASDEX Upgrade, COMPASS, ISTTOK, MAST, TJ-K, RFX, H-1 Heliac, IR-T1, GOLEM as well as low temperature devices as DC cylindrical magnetron in Prague and linear magnetized plasma devices in Nancy and Ljubljana.

Principle

If a Langmuir probe (electrode) is inserted into a plasma, its potential is not equal to the plasma potential V_{fl}. because a Debye sheath forms, but instead to a floating potential. The difference with the plasma potential is given by the electron temperature T_e:

$$\Phi - V_{fl} = \alpha * T_e$$

where the coefficient α is given by the ratio of the electron and ion saturation current density $\left(j_e^{sat} \, and \, j_e^{sat} \right)$ and collecting areas for electrons and ions $\left(A_e \, and \, A_i \right)$:

$$\alpha = ln\left(\frac{A_e j_e^{sat}}{A_i j_i^{sat}} \right) = ln(R)$$

The ball-pen probe modifies the collecting areas for electrons and ions in such a way that the ratio R is equal to one. Consequently, and the floating potential of the ball-pen probe becomes equal to the plasma potential regardless of the electron temperature:

$$V_{fl} = \Phi$$

Design and Calibration

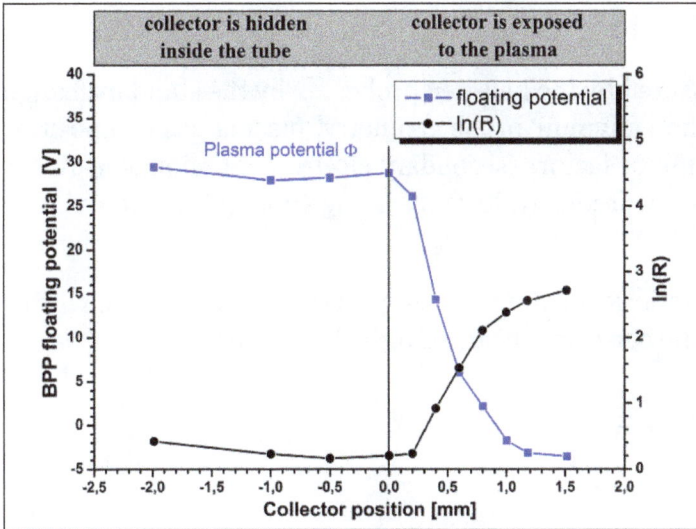

Potential and ln(R) of the ball-pen probe
for different positions of the collector.

A ball-pen probe consists of a conically shaped collector (non-magnetic stainless steel, tungsten, copper, molybdenum), which is shielded by an insulating tube (boron nitride, Alumina). The collector is fully shielded and the whole probe head is placed perpendicular to magnetic field lines.

When the collector slides within the shield, the ratio R varies, and can be set to 1. The adequate retraction length strongly depends on the magnetic field's value. The collector retraction should be roughly below the ion's Larmor radius. Calibrating the proper position of the collector can be done in two different ways:

1. The ball-pen probe collector is biased by a low-frequency voltage that provides the I-V characteristics and obtain the saturation current of electrons and ions. The collector is then retracted until the I-V characteristics becomes symmetric. In this case, the ratio R is close to unity, though not exactly. If the probe is retracted deeper, the I-V characteristics remain symmetric.

2. The ball-pen probe collector potential is left floating, and the collector is retracted until its potential saturates. The resulting potential is above the Langmuir probe potential.

Electron Temperature Measurements

Using two measurements of the plasma potential with probes whose coefficient α differ, it is possible to retrieve the electron temperature passively (without any input voltage or current). Using a Langmuir probe (with a non-negligible) and a ball-point probe (whose associated is close to zero) the electron temperature is given by:

$$T_e = \frac{\Phi - V_{fl}}{\alpha}$$

where Φ is measured by the ball-pen probe, V_{fl} by the standard Langmuir probe, and α is given by the Langmuir probe geometry, plasma gas composition, the magnetic field, and other minor factors (secondary electron emission, sheath expansion, etc.) It can be calculated theoretically, its value being about 3 for a non-magnetized hydrogen plasma.

In practice, the ratio R for the ball-pen probe is not exactly equal to one, so that the coefficient must be corrected by an empirical value for:

$$T_e = \frac{\Phi_{BPP} - V_{fl}}{\bar{\alpha}},$$

where:

$$\bar{\alpha} = \alpha - ln(R).$$

Faraday Cup

A Faraday cup is a metal (conductive) cup designed to catch charged particles in vacuum. The resulting current can be measured and used to determine the number of ions or electrons hitting the cup. The Faraday cup is named after Michael Faraday who first theorized ions around 1830.

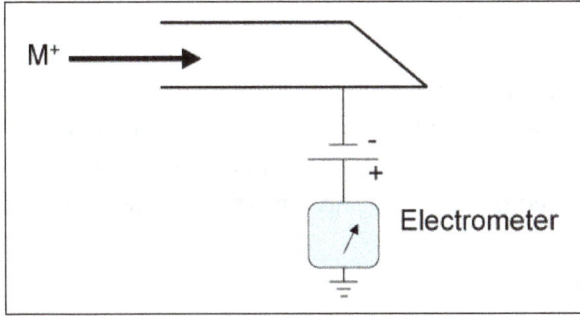

Schematic diagram of a Faraday cup.

It is installed in Space probes (Voyager 1, & 2, Parker Solar Probe, etc), or used in Mass spectrometry, and others.

Principle of Operation

When a beam or packet of ions hits the metal it gains a small net charge while the ions are neutralized. The metal can then be discharged to measure a small current proportional to the number of impinging ions. Essentially the Faraday cup is part of a circuit where ions are the charge carriers in vacuum and the Faraday cup is the interface to the solid metal where electrons act as the charge carriers (as in most circuits). By measuring the electric current (the number of electrons flowing through the circuit per second) in the metal part of the circuit the number of charges being carried by the ions in the vacuum part of the circuit can be determined. For a continuous beam of ions (each with a single charge):

$$\frac{N}{t} = \frac{I}{e}$$

where N is the number of ions observed in a time t (in seconds), I is the measured current (in amperes) and e is the elementary charge (about 1.60×10^{-19} C). Thus, a measured current of one nanoamp (10^{-9} A) corresponds to about 6 billion ions striking the faraday cup each second.

Faraday cup with an electron-suppressor plate in front.

Similarly, a Faraday cup can act as a collector for electrons in a vacuum (for instance from an electron beam). In this case electrons simply hit the metal plate/cup and a current is produced. Faraday cups are not as sensitive as electron multiplier detectors, but are highly regarded for accuracy because of the direct relation between the measured current and number of ions.This device is considered a universal charge detector because of its independence from the energy, mass, chemistry, etc. of the analyte.

Plasma Diagnostics

The Faraday cup utilizes a physical principle according to which the electrical charges delivered to the inner surface of a hollow conductor are redistributed around its outer surface due to mutual self-repelling of charges of the same sign – a phenomenon discovered by Faraday.

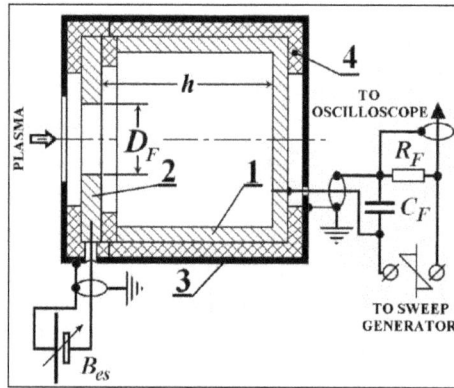

Faraday cup for plasma diagnostics.

The conventional Faraday cup is applied for measurements of ion (or electron) flows from plasma boundaries and comprises a metallic cylindrical receiver-cup – 1 closed with, and insulated from, a washer-type metallic electron-suppressor lid – 2 provided with the round axial through enter-hollow of an aperture with a surface area $S_F = \pi D_F^2 / 4$. Both the receiver cup and the electron-suppressor lid are enveloped in, and insulated from, a grounded cylindrical shield – 3 having an axial round hole coinciding with the hole in the electron-suppressor lid – 2. The electron-suppressor lid is connected by 50 Ω RF cable with the source B_{es} of variable DC voltage U_{es}. The receiver-cup is connected by 50 Ω RF cable through the load resistor R_F with a sweep generator producing saw-type pulses $U_g(t)$. Electric capacity C_F is formed of the capacity of the receiver-cup – 1 to the grounded shield – 3 and the capacity of the RF cable. The signal from R_F enables an observer to acquire an I-V characteristic of the Faraday cup by oscilloscope Proper operating conditions: $h \geq D_F$ (due to possible potential sag) and $h \ll \lambda_i$, where λ_i is the ion free path. Signal from R_F is the Faraday cup I-V characteristic which can be observed and memorized by oscilloscope:

$$i_\Sigma(U_g) = i_i(U_g) - C_F \frac{dU_g}{dt}.$$

In figure – cup-receiver, metal (stainless steel). 2 – electron-suppressor lid, metal (stainless steel). 3 – grounded shield, metal (stainless steel). 4 – insulator (teflon, ceramic). C_F – capacity of Faraday cup. R_F – load resistor.

Thus we measure the sum i_Σ of the electric currents through the load resistor $R_F i_i$ (Faraday cup current) plus the current $i_c(U_g) = -C_F(dU_g / dt)$ induced through the capacitor C_F by the saw-type voltage U_g of the sweep-generator: The current component $i_i(U_g)$ can be measured at the absence of the ion flow and can be subtracted further from the total current U_g measured with plasma to obtain the actual Faraday cup I-V characteristic $i_i(U_g)$ for processing. All of the Faraday cup elements and their assembly that interact with plasma are fabricated usually of temperature-resistant materials (often these are stainless steel and teflon or ceramic for insulators). For processing of the Faraday cup I-V characteristic, we are going to assume that the Faraday cup is installed far enough away from an investigated plasma source where the flow of ions could be considered as the flow of particles with parallel velocities directed exactly along the Faraday cup axis. In this case, the elementary particle current di_i corresponding to the ion density differential $dn(v)$ in the range of velocities between v and $v + dv$ of ions flowing in through operating aperture S_F of the electron-suppressor can be written in the form,

$$di_i = eZ_i S_F v dn(v)$$

where,

$$dn(v) = nf(v)dv$$

e is elementary electric charge, Z_i is the ion charge state, and $f(v)$ is the one-dimensional distribution function of ions over velocity v. Therefore, the ion current at the ion-decelerating voltage U_g of the Faraday cup can be calculated by integrating eq. $di_i = eZ_i S_F v dn(v)$ after substituting in it eq. above,

$$i_i(U_g) = eZ_i n_i S_F \int_{\sqrt{2eZ_i U_g / M_i}}^{\infty} f(v)v dv,$$

where the lower integration limit is defined from the obvious equation $M_i v_{i,s}^2 / 2 = eZ_i U_g$ where $v_{i,s}$ is the velocity of the ion stopped by the decelerating potential U_g, and M_i is the ion mass. Thus eq. above represents the I-V characteristic of the Faraday cup. Differentiating eq. above with respect to U_g, one can obtain the relation,

$$\frac{di_i(U_g)}{dU_g} = -en_i S_F \frac{eZ_i}{M_i} f\left(\sqrt{2eZ_i U_g / M_i}\right)$$

where the value $-n_i S_F (eZ_i / M_i) = C_i$ is an invariable constant for each measurement. Therefore, the average velocity $\langle v_i \rangle$ of ions arriving into the Faraday cup and their average energy $\langle \mathcal{E}_i \rangle$ can be calculated (under the assumption that we operate with a single type of ion) by the expressions,

$$\langle v_i \rangle = 1.389 \times 10^6 \sqrt{\frac{Z_i}{M_A}} \int_0^\infty i_i'(U_g) dU_g \left(\int_0^\infty \frac{i_i'}{\sqrt{U_g}} dU_g \right)^{-1} [cm/s],$$

$$\langle \mathcal{E}_i \rangle = \int_0^\infty i_i'(U_g) \sqrt{U_g} dU_g \left(\int_0^\infty \frac{i_i'}{\sqrt{U_g}} dU_g \right)^{-1} [eV],$$

where M_A is the ion mass in atomic units. The ion concentration n_i in the ion flow at the Faraday cup vicinity can be calculated by the formula,

$$n_i = \frac{i_i(0)}{eZ_i \langle v_i \rangle S_F}$$

which follows from eq. $i_i(U_g) = eZ_i n_i S_F \int_{\sqrt{2eZ_i U_g / M_i}}^\infty f(v) v dv$, at,

$$\int_0^\infty f(v) v dv = \langle v \rangle,$$

and from the conventional condition for distribution function normalizing,

$$\int_0^\infty f(v) dv = 1.$$

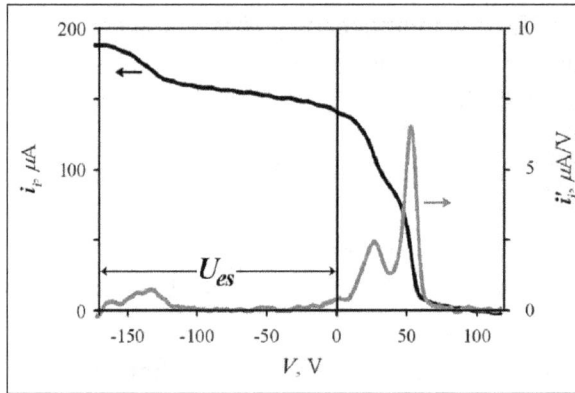

Faraday cup I-V characteristic.

The I-V characteristic $i_i(V)$ and its first derivative $i_i'(V)$ of the Faraday cup with $S_F = 0.5 cm^2$ installed at output of the Inductively coupled plasma source powered with RF 13.56 MHz and operating at 6 mTorr of H2. The value of the electron-suppressor voltage (accelerating the ions) was set experimentally at $U_{es} = -170V$, near the point of suppression of the secondary electron emission from the inner surface of the Faraday cup.

Error Sources

The counting of charges collected per unit time is impacted by two error sources: 1) the emission of low-energy secondary electrons from the surface struck by the incident charge and 2) backscattering (~180 degree scattering) of the incident particle, which causes it to leave the collecting surface, at least temporarily. Especially with electrons, it is fundamentally impossible to distinguish between a fresh new incident electron and one that has been backscattered or even a fast secondary electron.

SPECTRAL LINE RATIOS

The analysis of line intensity ratios is an important tool to obtain information about laboratory and space plasmas. In emission spectroscopy, the intensity of spectral lines can provide various information about the plasma (or gas) condition. It might be used to determine the temperature or density of the plasma. Since the measurement of an absolute intensity in an experiment can be challenging, the ratio of different spectral line intensities can be used to achieve information about the plasma, as well.

The emission intensity of an atomic transition from the upper state to the lower state is given in erg/cm^3s :

$$P_{u \to l} = N_u\, \hbar\omega_{u \to l}\, A_{u \to l},$$

where:

- N_u is the density of ions in the upper state,

- $\hbar\omega_{u \to l}$ is the energy of the emitted photon, which is the product of Planck's constant and the transition frequency,

- $A_{u \to l}$ is the Einstein coefficient for the specific transition.

The population of atomic states N is generally dependent on plasma temperature and density. Generally, the more hot and dense the plasma, the higher atomic states are populated. The observance or not-observance of spectral lines from certain ion species can, therefore, help to give a rough estimation of the plasma parameters.

More accurate results can be obtained by comparing line intensities:

$$\frac{P_{u_1 \to l_1}}{P_{u_2 \to l_2}} = \frac{N_{u_1}\omega_{u_1 \to l_1}A_{u_1 \to l_1}}{N_{u_2}\omega_{u_2 \to l_2}A_{u_2 \to l_2}},$$

The transition frequencies and the Einstein coefficients of transitions are well known and listed in various tables as in NIST Atomic Spectra Database. It is often that atomic modeling is required for determination of the population densities N_{u_1} and N_{u_2} as a function of density and temperature. While for the temperature determination of plasma in thermal equilibrium Saha's equation and Boltzmann's formula might be used, the density dependence usually requires atomic modeling.

PLASMA SPECTROSCOPY

Plasma spectroscopy is one of the most established and oldest diagnostic tools in astrophysics and plasma physics. Radiating atoms, molecules and their ions provide an insight into plasma processes and plasma parameters and offer the possibility of real-time observation. Emission spectra in the visible spectral range are easy to obtain with a quite simple and robust experimental set-up. The method itself is non-invasive, which means that the plasma is not affected. In addition, the presence of rf fields, magnetic fields, high potentials etc. does not disturb the recording of spectra. Also the set-up at the experiment is very simple: only diagnostic ports are necessary which provide a line-of-sight through the plasma. Thus plasma spectroscopy is an indispensable diagnostic technique in plasma processing and technology as well as in fundamental research. Although spectra are easily obtained, interpretation can be fairly complex, in particular, in low temperature, low pressure plasmas which are far from thermal equilibrium, i.e. non-equilibrium plasmas.

Radiation in the Visible Spectral Range

Electromagnetic waves extend over a wide wavelength range, from radio waves (kilometre) down to γ -rays (picometer). The visible range is only a very small part ranging from 380 to 780 nm by definition. However, common extensions are to the ultraviolet and the infrared resulting roughly in a range from 200 nm to 1 μm. From the experimental point of view this wavelength region is the first choice in plasma spectroscopy: air is transparent, quartz windows can be used and a variety of detectors and light sources are available. Below 200 nm quartz glass is no longer transparent and the oxygen in the air starts to absorb light resulting in the requirement of an evacuated light path. Above 1 μm the thermal background noise becomes stronger which can only be compensated for by the use of expensive detection equipment.

Radiation in the visible spectral range originates from atomic and molecular electronic transitions. Thus, the heavy particles of low temperature plasmas, the neutrals and their ions basically characterize the colour of a plasma: typically a helium plasma is pink, neon plasmas are red, nitrogen plasmas are orange and hydrogen are purple— these are first results of spectroscopic diagnostics using the human eye.

Emission and Absorption

In general, plasma spectroscopy is subdivided into two types of measurements: the passive method of emission spectroscopy and the active method of absorption spectroscopy.

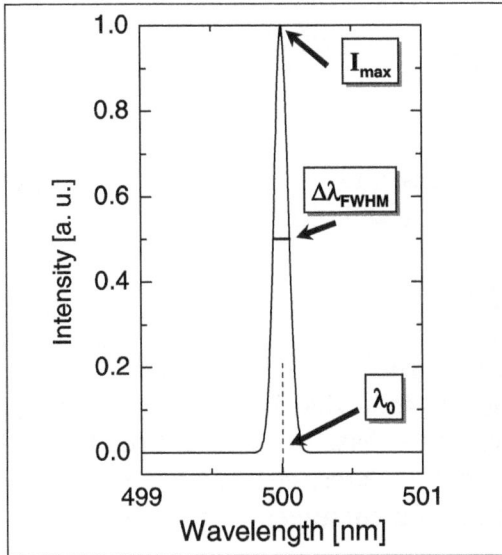

Line radiation and its characteristics.

In the case Of emission spectroscopy, light emitted from the plasma itself is recorded. Here, one of the basic underlying processes is the excitation of particles (atoms, molecules, ions) by electron impact from level q to level p and the decay into level k by spontaneous emission with the transition probability A_{pk} resulting in line emission ε_{pk}. In the case of absorption spectroscopy, the excitation from level q to level p takes place by a radiation field (i.e. by absorption with the transition probability B_{qp}) resulting in a weakening of the applied radiation field which is recorded. The intensity of emission is correlated with the particle density in the excited state $n(p)$, whereas the absorption signal correlates with the particle density in the lower state $n(q)$, which is in most cases the ground state. Thus, ground state particle densities are directly accessible by absorption spectroscopy; however, absorption techniques need much more experimental effort than emission spectroscopy. Since some principles of emission spectroscopy apply also to absorption and, since emission spectroscopy provides a variety of plasma parameters and is a passive and very convenient diagnostic tool these lecture notes are focused on emission spectroscopy. The two axes of a spectrum are the wavelength axis and the intensity axis as shown in figure. The central wavelength of line emission λ_0 is given by the photon energy $E = E_p E_k$ corresponding to the energy gap of the transition from level p with energy E_p to the energetically lower level k (Planck constant h, speed of light c):

$$\lambda_0 = hc / \left(E_p - E_k \right)$$

Since the energy of a transition is a characteristic of the particle species, the central wavelength is an identifier for the radiating particle, unless the wavelength is shifted by the Doppler effect. As a principle, the wavelength axis λ is easy to measure, to calibrate and to analyse. This changes to the opposite for the intensity axis. The line intensity is quantified by the line emission coefficient:

$$\varepsilon_{pk} = n(p) \, A_{pk} \frac{hc}{4\pi\lambda_0} = \int_{line} \varepsilon\lambda \, d\lambda$$

in units of W $(m^2 \, sr)^{-1}$, where 4π represents the solid angle $d\Omega$ (isotropic radiation), measured in steradian (sr). The line profile P_λ correlates the line emission coefficient with the spectral line emission coefficient ε_λ:

$$\varepsilon_\lambda = \varepsilon_{pk} P_\lambda \text{ with } \int_{line} P_\lambda \, d_\lambda = 1.$$

A characteristic of the line profile is the full width at half maximum (FWHM) of the intensity, $\Delta\lambda$FWHM, as indicated in figure. The line profile depends on the broadening mechanisms. In the case of Doppler broadening the profile is a Gaussian profile; the line width correlates with the particle temperature. A convenient alternative to the line emission coefficient is the absolute line intensity in units of photons $(m^3 \, s)^{-1}$:

$$I_{pk} = n(p) A_{pk}.$$

This relationship reveals that the line intensity depends only on the population density of the excited level $n(p)$ which, in turn, depends strongly on the plasma parameters $n(p) f(T_e, n_e, T_n, n_n, ...)$.

Atomic and Molecular Spectra

The atomic structure of atoms and molecules is commonly represented in an energy level diagram and is strongly related to emission (and absorption) spectra. The electronic energy levels of atoms and diatomic molecules have their spectroscopic notation:

$$n\ell^w \, {}^{2S+1}L_{L+S} \text{ and } n\ell^{w2S+1}\Lambda_\Lambda + \Sigma_{g,u}^{+,-},$$

respectively. n is the main quantum number, the angular momentum, w the number of electrons in the shell, S the spin, 2S +1 the multiplicity, L+S = J the total angular momentum. This represents the LS coupling which is valid for light atoms. Details of atomic structure can be found in the standard books. In case of diatomic molecules the projection of the corresponding vectors onto the molecular axis is important, indicated by Greek letters. +, − and g, u denote the symmetry of the electronic wave function. Optically allowed transitions follow the selection rules for dipole transitions which can be summarized as: L = 0, ±1, J = 0, ±1, S = 0 for atoms and = 0, u ↔ g for molecules. L = 0 or J = 0 transitions are not allowed if the angular momenta of both states involved are zero.

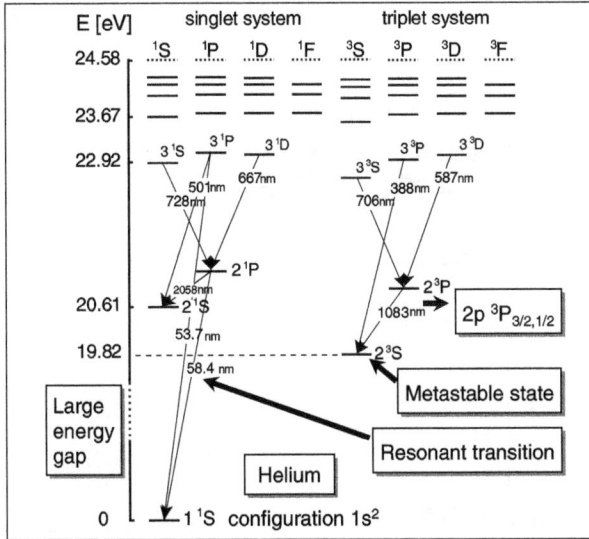

Example of an atomic energy level diagram.

In figure shows in the energy level diagram for helium which is a two electron system. The levels are separated into two multiplet systems: a singlet and a triplet system. Following Pauli's principle the spin of two electrons in the ground state is arranged anti-parallel resulting in the 1s 11S state. The fine structure is indicated only for the 23P state. Electronic states which cannot decay via radiative transitions have a long lifetime and are called metastable states (23S and 21S). Transitions which are linked directly to the ground state are called resonant transitions. The corresponding transition probability is high, hence the radiation is very intense. Due to the large energy gap these transitions are often in the vacuum ultraviolet (vuv) wavelength range. Optically allowed transitions with upper quantum number n 3 are indicated by an arrow and labelled with the corresponding wavelength. Radiation in the visible spectral range mostly originates from transitions between excited states.

The energy level diagram of a diatomic molecule with two electrons, i.e. molecular hydrogen, is shown schematically in figure. Again the two electrons cause a splitting into a singlet and a triplet system. In the united atom approximation for molecules a main quantum number can be assigned. In molecules the energy levels are usually abbreviated by upper and lower case letters, where X is the ground state (as a rule). The corresponding spectroscopic notation is shown in figure for the X and b states. Due to the additional degrees of freedom, each electronic state has vibrational levels (quantum number v) and each vibrational level has rotational levels (quantum number J) which appear with decreasing energy distances. The vibrational levels in the ground state are indicated in figure. A special feature in the energy level diagram of the hydrogen molecule is the repulsive state b 3 + u. Due to the repulsive potential curve, the energy range covers a few electronvolts. Molecules in this state eventually dissociate. Radiation in the visible spectral range correspond to an electronic transition without restrictions for the change in the vibrational quantum number and appear in the spectrum as vibrational bands. Each vibrational band has a rotational structure, where the rotational lines must follow the selection rules: J = 0, ±1, forming so-called P–, Q– and R–branches.

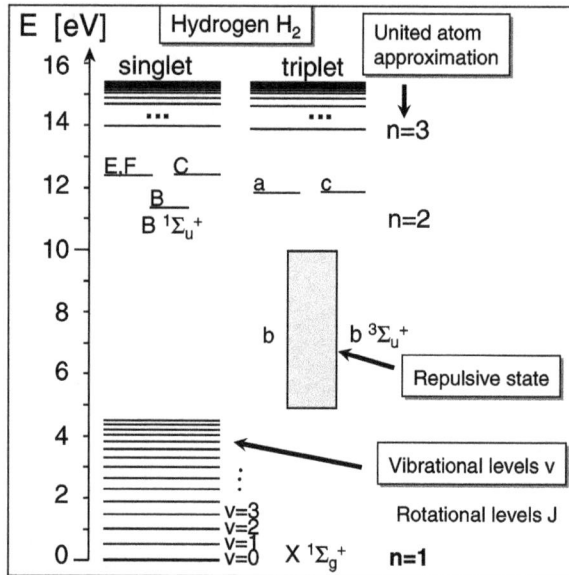

Example of an energy level diagram of a diatomic molecule.

An additional feature of diatomic molecules is the internuclear distance of the two nuclei. Thus, potential curves define the energy levels. This is shown in figure for the ground state and two electronically excited states of hydrogen together with vibrational levels, i.e. the vibrational eigenvalues and the corresponding vibrational wave functions. Under the assumption that the internuclear distance does not change in the electron impact excitation process and during the decay by spontaneous emission, excitation and de-excitation follow a vertical line in figure, which means that the Franck– Condon principle is valid. In other words, a vibrational population in the ground state is projected via the Franck– Condon factors into the electronically excited state. The Franck– Condon factor is defined as the overlap integral of two vibrational wave functions.

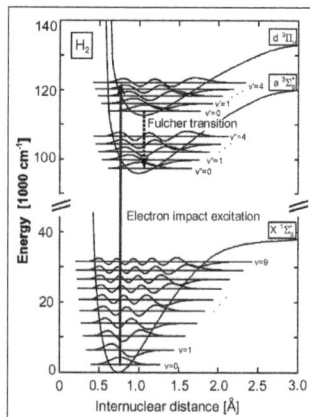

Molecular excitation and radiation according to the Franck–Condon principle for the ground state and two excited states of molecular hydrogen.

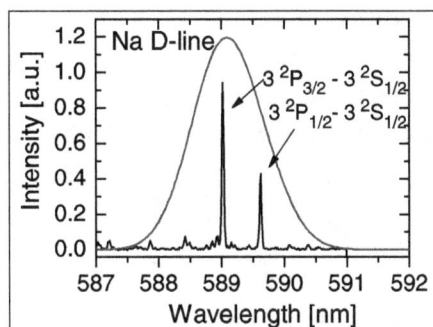

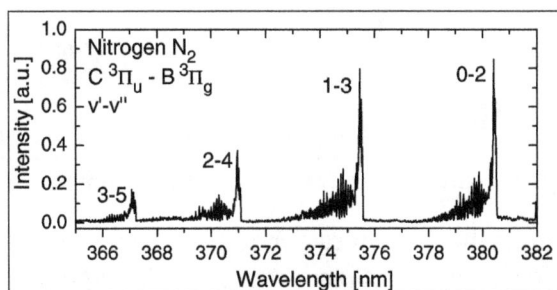

Atomic and molecular spectra: NaD-lines and vibrational
bands of the second positive system of N_2.

In addition for radiative transitions the electronic dipole transition momentum has to be taken into account. Thus, the Franck–Condon factors are replaced by vibrational transition probabilities and together with the vibrational population of the excited state determine the intensity of a vibrational band.

In figure shows the intense vibrational bands of molecular nitrogen. These vibrational bands v – v with v = 2 (v is the vibrational quantum number in the upper electronic state, v is the vibrational quantum number in the lower electronic state) correspond to the electronic transition C 3 u– B3 g, which is called the second positive system of nitrogen. The rotational structure of each vibrational band is observed clearly: however, the shape of the bands is determined by the spectral resolution of the spectroscopic system. The same applies to atomic spectra as shown in the upper part of figure 5 for sodium. The two narrow lines correspond to a recorded spectrum where the fine structure is resolved, whereas the broad line would be observed by a spectrometer with poor spectral resolution, a quantity.

Spectroscopic Systems

The choice of spectrographs, detectors and optics depends strongly on the purpose for which the diagnostic tool is to be used. Details of various spectroscopic systems and their components can be looked up in standard books about optics. The basic components of a spectrometer are: the entrance and exit slit, the grating as the dispersive element and the imaging mirrors, as illustrated in figure for the Czerny–Turner configuration. The exit slit is equipped with a detector. The light source, i.e. the plasma, is either imaged by an imaging optics onto the entrance slit or coupled by fibres to the slit. The latter is very convenient, particularly when direct access to the plasma light is difficult. The individual parts of the spectroscopic system determine the accessible wavelength range, the spectral resolution and the throughput of light.

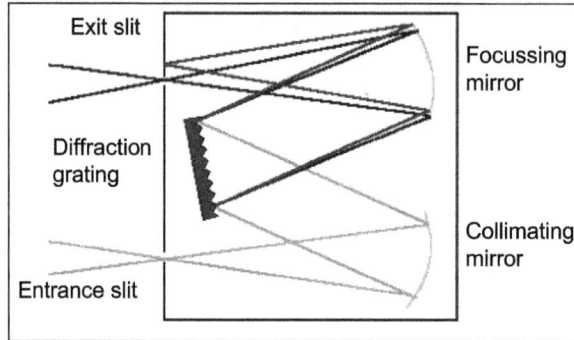

Monochromator in Czerny–Turner configuration.

The choice of the grating, which is characterized by the grooves per millimetre (lines/mm) is of importance for the spectral resolution. Special types of gratings such as Echelle gratings are optimized for high order diffractions resulting in a high spectral resolution. The blaze angle of a grating determines the wavelength range with highest reflection efficiency, i.e. the sensitivity of the grating.

The focal length of the spectrometer influences the spectral resolution and together with the size of the grating defines the aperture and thus the throughput of light. The width of the entrance slit is also of importance for light throughput, which means a larger entrance slit results in more intensity, with the drawback that the spectral resolution decreases.

At the exit either a photomultiplier is mounted behind the exit slit or a CCD (charge-coupled device) array is mounted directly at the image plane of the exit. In the first case, the width of the exit slit or, in the second case, the pixel size influence the spectral resolution of the system. The overall sensitivity of the system is strongly dominated by the type of detector: photomultipliers with different cathode coatings or CCD arrays with different sensor types (intensified, back-illuminated, etc.). Spectroscopic systems which use photomultipliers are scanning systems whereas systems with CCD arrays are capable of recording a specific wavelength range. Spatial resolution can be achieved by the choice of 2-dim detectors. Temporal resolution is completely determined by the detector: photomultipliers are usually very fast, whereas CCD arrays are limited by the exposure time and read-out time.

In the following some typical system set-ups are presented in order to give an overview of the different types of spectrometers and their typical application. For line monitoring, which means following the temporal behaviour of an emission line, pocket size survey spectrometers are very suitable. They have a poor spectral resolution $\lambda \approx 1\text{--}2$ nm but a good time resolution. The diagnostic technique itself is a simple one, providing information on plasma stability or changes in particle densities. Another typical system is a spectrometer with a focal length of 0.5–1 m ($\lambda \approx 40$ pm) and a grating with

1200 lines mm^{-1}. The optical components and the optics are very much improved; the time resolution depends on the detector. Using a 2–dim CCD camera reasonable spatial resolution can be achieved. This combination represents a flexible system and is a reasonable compromise between spectral resolution and temporal behaviour.

The next step is to deploy an Echelle spectrometer, which provides an excellent spectral resolution ($\partial\lambda$ 1–2 pm) by making use of the high orders of diffraction provided by the special Echelle grating. Typical applications are measurements of line profiles and line shifts. An important point in the choice of the spectroscopic system is the intensity of the light source. For example, measurements with an Echelle spectrometer require much more light than measurements with the survey spectrometer. However, this can be partly compensated by the choice of detector and exposure time.

Another important issue is the calibration of the spectroscopic system. One part is the calibration of the wavelength axis, which is an easy task, done by using spectral lamps (or the plasma itself) in combination with wavelength tables. For example a mercury–cadmium lamp can be used as the cadmium and mercury lines extend over a wide wavelength range. Groups of lines with various distances between each other (wavelength axis) are very well suited to determine the spectral resolution of the system and the apparatus profile. In this case, line broadening mechanisms must be excluded, for example, by using low pressure lamps.

Much more effort is needed for the calibration of the intensity axis, which can be either a relative or an absolute calibration. A relative calibration takes into account only the spectral sensitivity of the spectroscopic system along the wavelength axis. An absolute calibration provides in addition the conversion between measured signals (voltage or counts) in W/(m^2 sr) or to Photons/(m^3 s) according to equations $\varepsilon_{pk} = n(p)\, A_{pk}\dfrac{hc}{4\pi\lambda_0} = \int_{line} \varepsilon\lambda\, d\lambda$

and $I_{pk} = n(p)A_{pk}$. An absolute calibrated system provides calibrated spectra, which gives direct access to plasma parameters. Thus, the effort is compensated by an increase in information. For the intensity calibration light sources are required for which the spectral radiance is known. One of the most critical points in the calibration procedure is the imaging of the light source to the spectroscopic system. Here one must be very careful to conserve the solid angle which is often adjusted by using apertures.

Calibration standards in the visible spectral range, from 350 nm up to 900 nm, are tungsten ribbon lamps, providing black body radiation (grey emitter) and Ulbricht spheres (diffusive sources). For extensions to the uv range down to 200 nm, the continuum radiation of deuterium lamps is commonly used. Since such light sources must have high accuracy they are usually electrical stabilized but they alter in time. This means that their lifetime as calibration source is limited, which is less critical for relative calibration. Typical curves of a calibration procedure are shown in figure. The upper part is the spectral radiance of an Ulbricht sphere (provided by an enclosed data sheet) which is typically used for the calibration of a spectroscopic system with fibre optics.

The spectrum in the middle gives the measured intensity in units depending on the detector (e.g. CCD detector, counts per second). The recorded spectrum is already normalized to the exposure time. Dividing the spectral radiance curve by the recorded spectrum the conversion factor is obtained representing the spectral sensitivity of the system. As already mentioned, absolutely calibrated spectral systems are the most powerful tool in plasma diagnostics. Thus, the following sections discuss the analysis methods based on absolutely calibrated spectra. Nevertheless, some basic principles can be applied even if only relatively calibrated systems or systems without calibration are available.

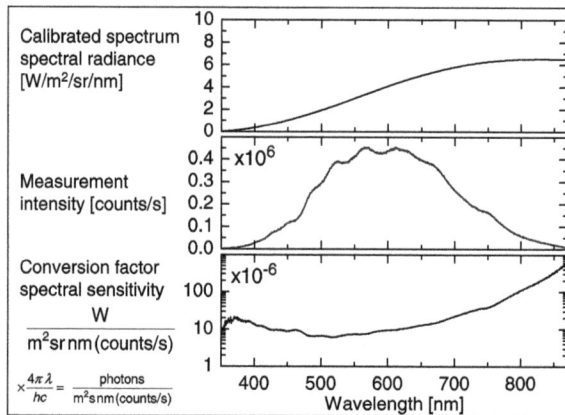

Example for the three steps needed to obtain a calibration curve.

Population Models

According to equation $I_{pk} = n(p)A_{pk}$, the absolute intensity of a transition is directly correlated to the population density in the excited state, the upper level. The population density of excited states is described by a Boltzmann distribution provided that the levels are in thermal equilibrium among each other. Since low temperature plasmas are non-equilibrium plasmas, which means they are far from (local) thermal equilibrium, the population density does not necessarily follow a Boltzmann distribution. As a consequence, the population in an excited state depends not only on the electron temperature but on a variety of plasma parameters: temperature and density of the electrons and the heavy particles, radiation field, etc. The dominant parameters are determined by the dominant plasma processes. Thus, population models are required which consider populating and depopulating processes for each individual level of a particle. An excellent overview of this topic is given in. It is obvious that for molecules where vibrational and rotational levels exist, the number of processes is incalculable and has to be reduced in some way.

Populating and Depopulating Processes

The electron impact excitation process is one of the most important processes. It increases the population of the upper level and decreases the population of the lower level. In turn electron impact de-excitation depopulates the upper level and populates the lower level. In a similar way, this principle applies to absorption and spontaneous emission for optically allowed transitions. Other population processes which couple with the particle in the next ionization stage are radiative or three-body recombination of the ion and the de-excitation by ionization.

Each process is described by its probability. In the case of spontaneous emission the probability is called the Einstein transition probability A_{ik}, where i labels the upper level and k the lower level. Collisional processes are generally described by cross sections or rate coefficients. The latter can be obtained from the convolution of the cross section with the corresponding energy distribution function of the impact particle. For an electron impact process the electron energy distribution is used, often described by a Maxwell distribution function. However, it should be kept in mind that this assumption is often not justified in low temperature plasmas. The upper part of figure shows Maxwellian electron energy distribution functions (EEDF) f (E) for two electron temperatures, Te 1.5 eV and Te 4.5 eV, and a typical cross section σ (E) (in arbitrary units) for an electron impact excitation process. The dashed area starting with the threshold energy of the excitation process (Ethr) indicates the part of the electrons which contributes to the rate coefficient. The lower part of figure shows the corresponding excitation rate coefficient Xexc(Te):

$$X_{exc}(T_e) = \int_{Ethr}^{\infty} \sigma\ (E)(2/m_e)^{1/2}\ \sqrt{E}\, f(E)\ dE$$

With

$$\int_0^{\infty} f(E)\ dE = 1.$$

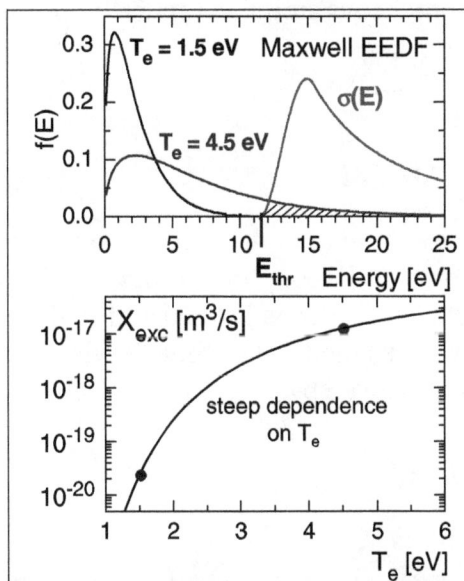

Convolution of a cross section with Maxwellian EEDFs
(Te = 1.5 eV and 4.5 eV) resulting in the excitation rate coefficient.

It is obvious that the rate coefficient Xexc(Te) shows a steep dependence on Te, in particular, at low electron temperatures, i.e. Te < Ethr. Since the part of the cross section with energies close to the threshold energy contributes most to the convolution, the accuracy of the rate coefficient depends strongly on the quality of the cross section in this energy region.

Corona Model

A simple approach to population densities in non-equilibrium plasmas is presented by the so-called corona model. The corona equilibrium is deduced from the solar corona where electron density is low ($\approx 10^{12}$ m^{-3}), electron temperature is high (≈ 100 eV) and where the radiation density is negligible. Due to the low electron density the probability of electron impact de-excitation processes is much lower than deexcitation by spontaneous emission and can be neglected. The electron temperature guarantees that the plasma is an ionizing plasma, i.e. recombination and thus populating processes from the next ionization stage do not play a role. Due to the insignificant radiation field, absorption is not important and excitation takes place only by electron impact collisions. Since these conditions are often fulfilled in low pressure, low temperature plasmas the usage of the corona model is a common method to deduce population (and ionization) equilibrium. However, the applicability has to be checked carefully in the individual case. These plasmas are characterized by a low degree of ionization. Each particle species (electrons, ions and neutrals) is characterized by its own temperature (under the assumption that a Maxwellian EEDF can be applied) and a gradual decrease is obtained: Te > Ti Tn.

The corona model assumes that upward transitions are only due to electron collisions while downward transitions occur only by radiative decay. Thus, in the simplest case, the population of an excited state p is balanced by electron impact excitation from the ground state q = 1 and decay by spontaneous emission (optically allowed transitions to level k):

$$n_1 n_e \, X_{1p}^{exc} \left(T_e \right) = n \left(p \right) \sum_t A_{pk} \ .$$

Collisional Radiative Model

A more general approach to population densities is to set up rate equations for each state of the particle together with the coupling to other particles, e.g. the next ionization stage. Since such a model balances the collisional and radiative processes the model is called a collisional radiative (CR) model. The time development of the population density of state p in a CR-model is given by:

$$\frac{dn\left(p \right)}{dt} = \sum_{k<p} n(k) n_e \, X_{kp}^{Xexc}$$

$$-n\left(p \right) \left[n_e (\sum_{k<p} X_{pk}^{de-exc} + \sum_{k<p} X_{pk}^{eXc} + s_p) + \sum_{k<p} A_{pk} \right]$$

$$+n(k) \sum_{k>p} (n_e \, X_{kp}^{de-exc} + A_{kp})$$

$$+ni \; ne \; \left(n_e \, \alpha_p + \beta_p\right).$$

The first term describes the electron impact excitation from energetically lower lying levels, $k < p$. Loss processes are electron impact de-excitation into energetically lower lying levels, $k < p$, electron impact excitation of energetically higher lying levels $k > p$, electron impact ionization rate coefficient (S(p)) and spontaneous emission. Next, electron impact processes and spontaneous emission from energetically higher lying levels $k > p$ is taken into account. The last two expressions describe population by three-body recombination and radiative recombination respectively. In addition, opacity, which means self-absorption of emission lines, may be important. A convenient method is to use the population escape factors which reduce the corresponding transition probabilities in equation above.

In the quasi-stationary treatment the time derivative is neglected (d n(p)/d t = o). However, transport effects may be of importance for the ground state and metastable states. This has to be taken into account by introducing diffusion or confinement times. According to the quasi-steadystate solution the set of coupled differential equations above is transformed into a set of coupled linear equations which depends on the ground state density and ion density. These equations are readily solved in the form,

$$n(p) = R_n(p) \, n_n \, n_e + R_i(p) \, n_i \, n_e$$

Rn(p) and Ri(p) are the so-called collisional–radiative coupling coefficients describing ground state and ionic population processes, respectively. In ionizing plasmas, the coupling to the ground state is of relevance, whereas in recombing plasmas the coupling to the ionic state dominates.

In contrast to the corona model, the population of an excited state in the CR model depends on more parameters than on the electron temperature only. The population coefficients depend on electron temperature and electron density. Further parameters can be temperature and density of other species such as neutrals. In general, the CR model closes the gap between the corona and Boltzmann equilibrium. Figure shows an example for population densities (normalized to the ground state) of the first electronically excited states of atomic hydrogen at T_e = 3 eV (ionizing plasma). The different regimes are indicated as well.

Since CR models depend strongly on the underlying data, it is obvious that the quality of results from CR models rely on the existence and quality of the cross sections (or rate coefficients). CR models are well established for atomic hydrogen, helium and argon which are elements with clear atomic structure. For molecules, CR models are scarce due to the manifold of energy levels to be considered and to the lack of data.

Preliminary models for molecular hydrogen and molecular nitrogen are now available. The complexity of such models requires comparisons with experimental results in a wide parameter range, i.e. validations of models by experimental results.

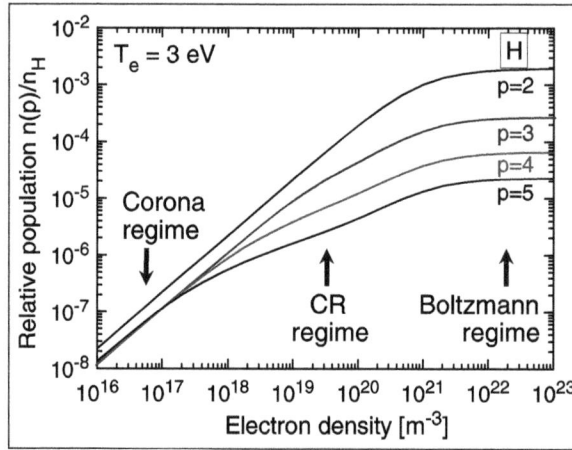

Results obtained from a CR model for atomic hydrogen.

The correlation between measured line emission and results from CR calculations are given by combining X_{pk}^{eff}. is the effective emission rate coefficient, depending not only on T_e but also on other plasma parameters such as n_e, T_n, n_n, etc.

$$\frac{I_{pk}}{n_n \, n_e} = \frac{n(p)}{n_n \, n_e} A_{pk} = \left(R_n(p) + R_i(p) n_i / n_n\right) A_{pk} = X_{pk}^{eff}.$$

References

- W. Amatucci; et al. (2001). "Contamination-free sounding rocket Langmuir probe". Review of Scientific Instruments. 72 (4): 2052–2057. Bibcode:2001RScI...72.2052A. doi:10.1063/1.1357234

- Ghoranneviss, M.; S. Meshkani (2016). "Techniques for improving plasma confinement in IR-T1 Tokamak". International Journal of Hydrogen Energy. 41 (29): 12555–12562. doi:10.1016/j.ijhydene.2016.03.075

- E. V. Shun'ko. (2009). Langmuir Probe in Theory and Practice. Universal Publishers, Boca Raton, Fl. 2008. p. 249. ISBN 978-1-59942-935-9

- Ralchenko, Yu. V.; Maron, Y. (2001). "Accelerated recombination due to resonant deexcitation of metastable states". J. Quant. Spectr. Rad. Transfer. 71 (2–6): 609–621. arXiv:physics/0105092. Bibcode:2001JQSRT..71..609R. CiteSeerX 10.1.1.74.3071. doi:10.1016/S0022-4073(01)00102-9

Plasma Processing

The plasma-based material processing technology which modifies the chemical and physical properties of a surface is referred to as plasma processing. A few of the techniques include plasma surface activation, plasma ashing, plasma cleaning, corona treatment, plasma etching, plasma functionalization, plasma polymerization, etc. All these diverse techniques of plasma processing have been carefully analyzed in this chapter.

Plasma processing technologies are of vital importance to several of the largest manufacturing industries in the world. Foremost among these industries is the electronics industry, in which plasma-based processes are indispensable for the manufacture of very large-scale integrated microelectronic circuits. Plasma processing of materials is also a critical technology in, for example, the aerospace, automotive, steel, biomedical, and toxic waste management industries. Most recently, plasma processing technology has been utilized increasingly in the emerging technologies of diamond film and superconducting film growth.

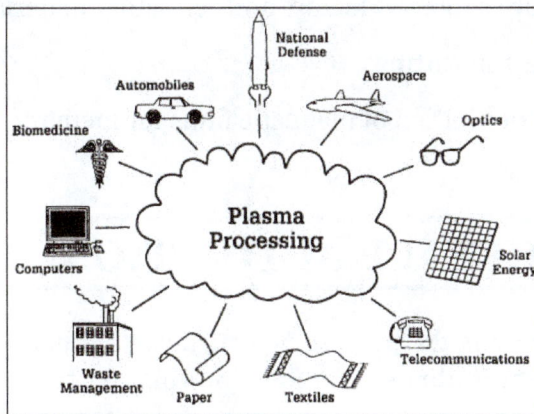

Plasma processing is a critical technology in many vital U.S. industries.

Plasmas used in materials processing cover a large portion of the low-energy density-temperature space.

Industrial Applications of Plasma Processing of Materials

The number of industrial applications of plasma-based systems for processing of materials and for surface modification is extensive, and many industries are impacted. Some of these processes and corresponding applications include:

- Plasma-controlled anisotropic etching in fabrication of microelectronic chips.

- Plasma deposition of silicon nitride for surface passivation and insulation.

- Surface oxidation used in fabrication of silicon-based microelectronic circuits.

- Plasma-enhanced chemical vapor deposition of amorphous silicon films used in solar cells.

- Plasma-surface treatment for improved film adhesion to polymer surfaces.

- Plasma nitriding, which is used to harden the surface of steel.

- Plasma-enhanced chemical vapor deposition and thermal plasma chemical vapor deposition of diamond thin films.

- Plasma spray deposition of ceramic or metal alloy coatings used for protection against wear or corrosion in aircraft and automotive engines.

- Plasma spray deposition of clearance control coatings.

- Plasma melting and refining of alloys.

- Plasma-assisted manufacture of optical fibers used in communications.

- Plasma synthesis of ultrapure powders used as ceramic precursors.

- Plasma spray deposition and thermal plasma chemical vapor deposition of high-temperature superconductors and refractory materials.

- Plasma welding and cutting.

- Plasma sputter deposition of magnetic films for memory devices.

PLASMA SURFACE ACTIVATION

Plasma surface activation is the process by which surface polymer functional groups are replaced with different atoms from ions in a plasma to increase surface energy. Plasma activation is often used to prepare a surface for bonding or printing.

Surface exposure to energetic species breaks down the polymer at the surface, creating free radicals. Plasma contains UV radiation at high levels, creating additional free radicals on the plastic or Teflon surface.

Free radicals quickly react with the material itself because they are unstable. This allows the surface to form stable covalent bonds, which are able to be printed on or bonded to.

Plasma technology is useful whenever one layer needs to be bonded to another layer. It simplifies manufacturing processes and improves the reliability and consistency of results.

Plasma Surface Activation.

Activation also improves the wettability of most surfaces, including aluminum and copper coatings. Most metal surfaces can improve their wettability with this process. Improved wettability is important for painting or printing with ink on plastic surfaces.

The most effective way to prepare a plastic or Teflon surface for printing, painting, or bonding is with plasma treatment.

Non-reactive or non-wettable surfaces may also be modified to obtain better bondability and adhesion properties. Materials coated or decorated without any form of surface activation will not adhere to the bond or the ink and thus be unreliable and messy.

Our plasma technology delivers the fastest, most economical, and most environmentally friendly solution available today.

Production timing is very important; the effects of the plasma are temporary, ranging from just a few hours to a few days. This allows plenty of time to complete the manufacturing process on the materials being treated if the manufacturing process is well planned.

Several mechanisms contribute to successful treatment depending on the materials treated and the gas used.

Gases used for Plasma Activation

Treating with oxygen gas has a significant etching effect by removing atoms from the surface one by one. This leaves a clean, bondable surface that will also accept ink or paint and create a permanent bond. If a metal surface is being treated, Argon gas is introduced with the Oxygen, to prevent oxidation of the treated surface.

Treating with oxygen will reverse the polarity of plastic materials and allow the polymers to receive new surface functionalities by increasing the surface tension and creating very small contact angles on the treated materials' surface.

There are numerous advantages to plasma processing as opposed to traditional wet-chemical activation methods. Most importantly, a second plasma treatment can be performed at any time with no degradation in the product being treated. This is not true of many treatment methods using harsh chemicals that could damage the material being treated.

Additionally, vacuum plasma is the greenest form of deep cleaning technologies in existence today. With no chemical waste, no danger to employees or facilities, and no chemicals to buy or store, being green has never been easier.

Plasma activation is most often used in the manufacturing of electronics, specifically in electronic sensors and connectors. It is useful prior to encapsulation because it strengthens the physical bond, and it ensures a reliable hermetic seal and reduces current leakage.

Plasma technology promotes increased flow of resins into plastic and polymer materials by raising the surface energy and enhancing the wettability of the treated products.

New chemical functional groups are formed with strong chemical bonds. Plasma surface activation allows more of the plastic to get into every crevice of the bonded surface. This results in improved bondability - up to 50 times the strength of a bond that is not treated with plasma activation.

PLASMA ASHING

In semiconductor manufacturing plasma ashing is the process of removing the photoresist (light sensitive coating) from an etched wafer. Using a plasma source, a monatomic (single atom) substance known as a reactive species is generated. Oxygen or fluorine are the most common reactive species. The reactive species combines with the photoresist to form ash which is removed with a vacuum pump.

Typically, monatomic oxygen plasma is created by exposing oxygen gas at a low pressure (O_2) to high power radio waves, which ionise it. This process is done under vacuum in order to create a plasma. As the plasma is formed, many free radicals are created which could damage the wafer. Newer, smaller circuitry is increasingly susceptible to these particles. Originally, plasma was generated in the process chamber, but as the need to get rid of free radicals has increased, many machines now use a downstream plasma configuration, where plasma is formed remotely and the desired particles are channeled to the wafer. This allows electrically charged particles time to recombine before they reach the wafer surface, and prevents damage to the wafer surface.

Two forms of plasma ashing are typically performed on wafers. High temperature ashing, or stripping, is performed to remove as much photo resist as possible, while the "descum" process is used to remove residual photo resist in trenches. The main difference between the two processes is the temperature the wafer is exposed to while in an ashing chamber.

Monatomic oxygen is electrically neutral and although it does recombine during the channeling, it does so at a slower rate than the positively or negatively charged free radicals, which attract one another. This means that when all of the free radicals have recombined, there is still a portion of the active species available for process. Because a large portion of the active species is lost to recombination, process times may take longer. To some extent, these longer process times can be mitigated by increasing the temperature of the reaction area.

PLASMA CLEANING

Plasma cleaning is the removal of impurities and contaminants from surfaces through the use of an energetic plasma or dielectric barrier discharge (DBD) plasma created from gaseous species. Gases such as argon and oxygen, as well as mixtures such as air and hydrogen/nitrogen are used. The plasma is created by using high frequency voltages (typically kHz to >MHz) to ionise the low pressure gas (typically around 1/1000 atmospheric pressure), although atmospheric pressure plasmas are now also common.

Methods

In plasma, gas atoms are excited to higher energy states and also ionized. As the atoms and molecules 'relax' to their normal, lower energy states they release a photon of light, this results in the characteristic "glow" or light associated with plasma. Different gases give different colors. For example, oxygen plasma emits a light blue color.

A plasma's activated species include atoms, molecules, ions, electrons, free radicals, metastables, and photons in the short wave ultraviolet (vacuum UV, or VUV for short) range. This mixture then interacts with any surface placed in the plasma.

If the gas used is oxygen, the plasma is an effective, economical, environmentally safe method for critical cleaning. The VUV energy is very effective in the breaking of most organic bonds (i.e., C–H, C–C, C=C, C–O, and C–N) of surface contaminants. This helps to break apart high molecular weight contaminants. A second cleaning action is carried out by the oxygen species created in the plasma (O_2^+, O_2^-, O_3, O, O^+, O^-, ionised ozone, metastable excited oxygen, and free electrons). These species react with organic contaminants to form H_2O, CO, CO_2, and lower molecular weight hydrocarbons. These compounds have relatively high vapor pressures and are evacuated from the chamber during processing. The resulting surface is ultra-clean. In figure, a relative content of carbon over material depth is shown before and after cleaning with excited oxygen.

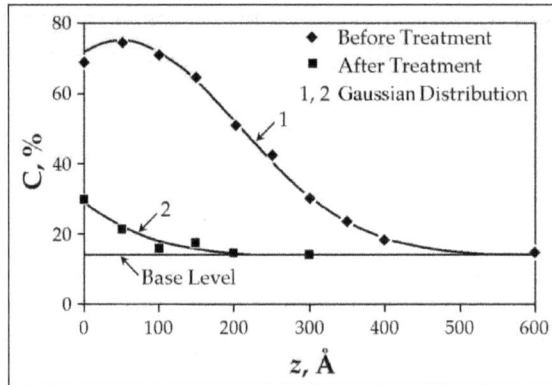

Content of carbon over material depth z: before a sample
treatment - diamond points and after the treatment during 1 s. - square points.

If the part to be treated consists of easily oxidized materials such as silver or copper, inert gases such as argon or helium are used instead. The plasma activated atoms and ions behave like a molecular sandblast and can break down organic contaminants. These contaminants are again vapourised and evacuated from the chamber during processing.

Most of these by-products are small quantities of gases such as carbon dioxide and water vapor with trace amounts of carbon monoxide and other hydrocarbons.

Whether or not organic removal is complete can be assessed with contact angle measurements. When an organic contaminant is present, the contact angle of water with the device will be high. After the removal of the contaminant, the contact angle will be reduced to that characteristic of contact with the pure substrate.

If a surface to be treated is coated with a patterned conductive layer (metal, ITO) deposited on it, treatment by a direct contact with plasma (capable for contraction to microarcs) could be destructive. In this case, cleaning by neutral atoms excited in plasma to metastable state can be applied. Results of the same applications to surfaces of glass samples coated with Cr and ITO layers are shown in figure.

After treatment, the contact angle of a water droplet is decreased becoming less than its value on the untreated surface. The relaxation curve for droplet footprint is shown for glass sample. A photograph of the same droplet on the untreated surface is shown in figure. Surface relaxation time corresponding to a data shown in figure is about 4 hours.

Plasma ashing is the process in which plasma cleaning is performed for the sole purpose of removing carbon. Plasma ashing is always done with O_2 gas.

Applications

Plasma cleaning is often required for the removal of contaminants from surfaces before they can be used in a manufacturing process. Plasma cleaning can be applied to an array of materials along with surfaces with complex geometries. Plasma cleaning has the capabilities to effectively remove all organic contaminations from surfaces through the process of a chemical reaction (air plasma) or physical ablation (Ar plasma/Argon plasma).

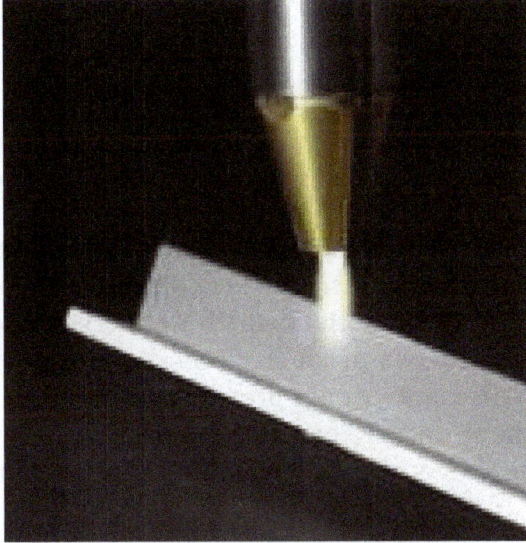

Plasma beam cleaning a metal surface.

CORONA TREATMENT

Verner Eisby, a Danish engineer, the
inventor of corona treatment.

Corona treatment (sometimes referred to as air plasma) is a surface modification technique that uses a low temperature corona discharge plasma to impart changes in the properties of a surface. The corona plasma is generated by the application of high voltage to an electrode that has a sharp tip. The plasma forms at the tip. A linear array of electrodes is often used to create a curtain of corona plasma. Materials such as plastics, cloth, or paper may be passed through the corona plasma curtain in order to change the surface energy of the material. All materials have an inherent surface energy. Surface treatment systems are available for virtually any surface format including dimensional objects, sheets and roll goods that are handled in a web format. Corona treatment is a widely used surface treatment method in the plastic film, extrusion, and converting industries.

The corona treatment was invented by the Danish engineer Verner Eisby in 1951. Verner had been asked by one of his customers if he could find a solution which would make it possible to print on plastic. Verner found that there were already a couple of ways to accomplish this. One was a gas flame method and the other was a spark generating method, both of which were crude and uncontrollable and did not produce a homogeneous product. Verner came up with the theory that a high frequency corona discharge would provide both a more efficient and controllable method to treat the surface. Exhaustive experiments proved him to be correct. Verner's company, Vetaphone, obtained patent rights for the new corona treatment system.

Materials

Many plastics, such as polyethylene and polypropylene, have chemically inert and non-porous surfaces with low surface tensions causing them to be non-receptive to bonding with printing inks, coatings, and adhesives. Although results are invisible to the naked eye, surface treating modifies surfaces to improve adhesion.

Polyethylene, polypropylene, nylon, vinyl, PVC, PET, metalized surfaces, foils, paper, and paperboard stocks are commonly treated by this method. It is safe, economical, and delivers high line speed throughput. Corona treatment is also suitable for the treatment of injection and blow molded parts, and is capable of treating multiple surfaces and difficult parts with a single pass.

Equipment

Corona discharge equipment consists of a high-frequency power generator, a high-voltage transformer, a stationary electrode, and a treater ground roll. Standard utility electrical power is converted into higher frequency power which is then supplied to the treater station. The treater station applies this power through ceramic or metal electrodes over an air gap onto the material's surface.

Two basic corona treater stations are used in extrusion coating applications—*Bare Roll* and *Covered Roll*. On a bare roll treater station, the dielectric encapsulates the electrode. On a covered roll station, it encapsulates the treater base roll. The treater consists of an electrode and a base roll in both stations. In theory a covered roll treater is generally used to treat non-conductive webs, and a Bare Roll treater is used to treat conductive webs. However, manufacturers who treat a variety of substrates on the same production line may choose to use a Bare Roll treater.

Pre-treatment

Many substrates provide a better bonding surface when they are treated at the time they are produced. This is called "pre-treatment." The effects of corona treatment diminish over time. Therefore many surfaces will require a second "bump" treatment at the time they are converted to ensure bonding with printing inks, coatings, and adhesives.

Other Technologies

Other technologies used for surface treatment include in-line atmospheric (air) plasma, flame plasma, and chemical plasma systems.

Atmospheric Plasma Treatment

Atmospheric-pressure plasma treatment is very similar to corona treatment but there are a few differences between them. Both treatments may use one or more high voltage electrodes which charge the surrounding blown gas molecules and ionizes them. However in atmospheric plasma systems, the overall plasma density is much greater which enhances the rate and degree to which the ionized molecules are incorporated onto a materials' surface. An increased rate of ion bombardment occurs which may result in stronger material bonding traits depending on the gas molecules used in the process. Atmospheric plasma treatment technology also eliminates a possibility of treatment on a material's non-treated side; also known as backside treatment.

Flame Plasma

Flame plasma treaters generate more heat than other treating processes, but materials treated through this method tend to have a longer shelf-life. These plasma systems are different from air plasma systems because flame plasma occurs when flammable gas and surrounding air are combusted into an intense blue flame. Objects' surfaces are polarized from the flame plasma affecting the distribution of the surface's electrons in an oxidation form. This treatment requires higher temperatures so many of the materials that are treated with a flame plasma can be damaged.

Chemical Plasma

Chemical plasma is based on the combination of air plasma and flame plasma. Much like air plasma, chemical plasma fields are generated from electrically charged air. But, instead of air, chemical plasma relies on a mixture of other gases depositing various chemical groups onto the treated surface.

PLASMA ETCHING

Plasma etching is a form of plasma processing used to fabricate integrated circuits. It involves a high-speed stream of glow discharge (plasma) of an appropriate gas mixture being shot (in pulses) at a sample. The plasma source, known as etch species, can be either charged (ions) or neutral (atoms and radicals). During the process, the plasma generates volatile etch products at room temperature from the chemical reactions between the elements of the material etched and the reactive species generated by the plasma. Eventually the atoms of the shot element embed themselves at or just below the surface of the target, thus modifying the physical properties of the target.

Mechanisms

Plasma Generation

A plasma is a high energetic condition in which a lot of processes can occur. These processes happen because of electrons and atoms. To form the plasma electrons have to be accelerated to gain energy. Highly energetic electrons transfer the energy to atoms by collisions. Three different processes can occur because of this collisions:

- Excitation.

- Dissociation.

- Ionization.

Different species are present in the plasma such as electrons, ions, radicals, and neutral particles. Those species are interacting with each other constantly. Plasma etching can be divided into two main types of interaction:

- Generation of chemical species.

- Interaction with the surrounding surfaces.

Without a plasma, all those processes would occur at a higher temperature. There are different ways to change the plasma chemistry and get different kinds of plasma etching or plasma depositions. One of the excitation techniques to form a plasma is by using RF excitation of a power source of 13.56 MHz.

The mode of operation of the plasma system will change if the operating pressure changes. Also, it is different for different structures of the reaction chamber. In the simple case, the electrode structure is symmetrical, and the sample is placed upon the grounded electrode.

Influences on the Process

The key to develop successful complex etching processes is to find the appropriate gas etch chemistry that will form volatile products with the material to be etched as shown in table. For some difficult materials (such as magnetic materials), the volatility can only be obtained when the wafer temperature is increased. The main factors that influence the plasma process:

- Electron source.

- Pressure.

- Gas species.

- Vacuum.

Surface Interaction

The reaction of the products depend on the likelihood of dissimilar atoms, photons, or radicals reacting to form chemical compounds. The temperature of the surface also affects the reaction of products. Adsorption happens when a substance is able to gather and reach the surface in a condensed layer, ranging in thickness (usually a thin, oxidized layer.) Volatile products desorb in the plasma phase and help the plasma etching process as the material interacts with the sample's walls. If the products are not volatile, a thin film will form at the surface of the material. Different principles that affect a sample's ability for plasma etching:

Table: Halogen-.hydride- and methyl-compounds and their volatility for elements and materials of interest in micro- and nano-technology applications.

Elements	Fluorides	Boiling temperature {°C}	Chlorides	Boiling temperature (°C)	Bromides	Boiling temperanire {°C}	Hydrides. trimethyls	Boiling temperature (°C)
Al	AlF_3	1297(subl.)	$AlCl_3$	178 (subl.)	$AlBr_3$	263		
As	AsF_3	−63	$AsCl_3$	130.2	$AsBr_3$	221	AsH_3	−55
	AsF_5	−53			$AsBr_5$			
C	CF_4	−128	CCl_4	77	CBr_4	189	CH_4	−164
C1	CrF_2	>1300	CrO_2Cl_2	117	$CrBr_2$	842		
Cu	CuF	1100(subl.)	CuCl	1490	CuBr	1345		
Ga	CuF_2	950	$CuCl_2$	993			CuH	55 60
Ge	GeF_4	1000	$GaCl_3$	201.3	$GaBr_3$	278.8	$Ga(CH_3)_3$	134
In	InF_3	−37 (subl.)	$GeCl_4$	84	$GeBr_4$	186.5	GeH_4	−88.5
Mo	MoF_5	> 1200	$InCl_3$	300 (subl.)			$In(CH_3)_3$	55.7
	MoF_6	213.6	$MoCl_5$	268				
	MoO_2F_2	35	$MoOCl_3$	100 (subl.)				
	$MoOF_4$	270 (subl.)						

P	PF_3	180	PCl_3					
	PF_5	−101.5	PCl_5	75	PBr_3	172.9	PH_3	−87.7
Si	SiF_4	−86	$SiCl_4$	162 (subl.)	PBr_5	106		
Ta	TaF_5	229.5	$TaCl_5$	57.6	$SiBr_4$	154	SiH_4	−111.8
Ti	TiF_4	284 (subl.)	$TeCl_4$	242	$TaBr_5$	348.8		
W	WF_6	17.5	WCl_6	136.4	$TiBr_4$	230		
	WOF_4	187.5	WCl_5	346.7				
			$WOCL_4$	275.6	WBr_5	333		
				227.5	$WoBr_4$	327		

- Volatility.
- Adsorption.
- Chemical Affinity.
- Ion-bombarding.
- Sputtering.

Plasma etching can change the surface contact angles, such as hydrophilic to hydrophobic, or vice versa. Argon plasma etching has reported to enhance contact angle from 52 deg to 68 deg, and, Oxygen plasma etching to reduce contact angle from 52 deg to 19 deg for CFRP composites for bone plate applications. Plasma etching has been reported to reduce the surface roughness from hundreds of nanometers to as much lower as 3 nm for metals.

Types

Pressure influences the plasma etching process. For plasma etching to happen, the chamber has to be under low pressure, less than 100 Pa. In order to generate low-pressure plasma, the gas has to be ionized. The ionization happens by a glow charge. Those excitations happen by an external source, which can deliver up to 30 kW and frequencies from 50 Hz (dc) over 5–10 Hz (pulsed dc) to radio and microwave frequency (MHz-GHz).

Microwave Plasma Etching

Microwave etching happens with an excitation sources in the microwave frequency, so between MHz and GHz. One example of plasma etching is shown here.

The microwave operates at 2.45 GHz. This frequency is generated by a magnetron and discharges through a rectangular and a round waveguide. The discharge area is in a quartz tube with an inner diameter of 66mm. Two coils and a permanent magnet are wrapped around the quartz tube to create a magnetic field which directs the plasma.

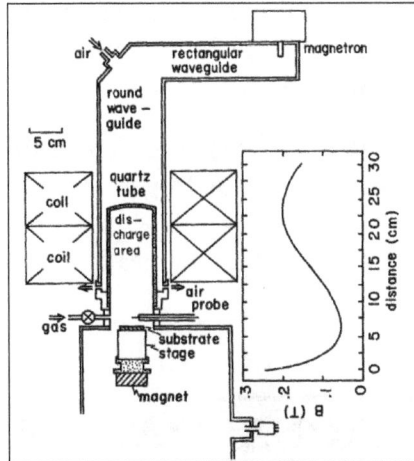

A microwave plasma etching apparatus.

Hydrogen Plasma Etching

One form to use gas as plasma etching is hydrogen plasma etching. Therefore, an experimental apparatus like this can be used:

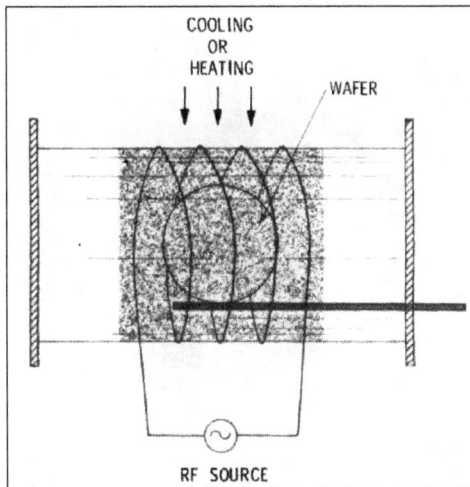

A quartz tube with an rf excitation of 30 MHz is shown.

It is coupled with a coil around the tube with a power density of 2-10 W/cm^3. The gas species is H$_2$ gas in the chamber. The range of the gas pressure is 100-300 um.

Applications

Plasma etching is currently used to process semiconducting materials for their use in the fabrication of electronics. Small features can be etched into the surface of the semiconducting material in order to be more efficient or enhance certain properties when used in electronic devices. For example, plasma etching can be used to create deep trenches on the surface of silicon for uses in microelectromechanical systems. This application suggests that plasma etching also has the potential to play a major role in the production of microelectronics. Similarly, research is currently being done on how the process can be adjusted to the nanometer scale.

Hydrogen plasma etching, in particular, has other interesting applications. When used in the process of etching semiconductors, hydrogen plasma etching has been shown to be effective in removing portions of native oxides found on the surface. Hydrogen plasma etching also tends to leave a clean and chemically balanced surface, which is ideal for a number of applications.

PLASMA FUNCTIONALIZATION

Plasma functionalization is the process by which a bondable surface is treated to improve its adhesive properties, including bondability or paintability.

Oxygen plasma is used in most applications, allowing printing or bonding to a surface that would otherwise be difficult if not impossible to print on or bond to.

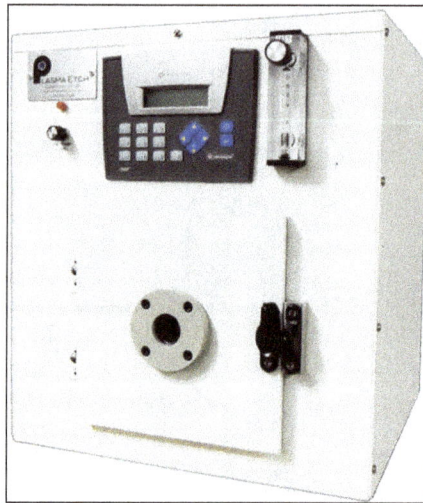

PE-25 Entry Level Plasma System.

If there is metal in or on the products being treated, argon is often added to the input gas mixture to prevent oxidation of any metal surfaces inside the chamber.

The treatment removes organic contaminantes and leaves a free radical on the surface being functionalized, which allows inks and glues to be accepted. These free radicals are chemically unstable and searching for an extra electron, which is why they bond so well.

Low frequency plasma is often used to functionalize surfaces, although high frequency plasma also works well.

Printing

Plasma functionalization leaves chemically unstable free radicals on the surface of most materials, allowing printing on a shiny surface that would otherwise smear or resist the ink. These surfaces include plastics and rubber products. The plasma process allows the surface to accept the ink.

Bonding

The PE-100 is our most popular system used for plasma functionalization. The convenient chamber size, along with the availability of high frequency RF power, combine to create a production capable unit small enough to fit nearly anywhere in your lab or production work space.

PLASMA POLYMERIZATION

Plasma polymerization (or glow discharge polymerization) uses plasma sources to generate a gas discharge that provides energy to activate or fragment gaseous or liquid monomer, often containing a vinyl group, in order to initiate polymerization. Polymers formed from this technique are generally highly branched and highly cross-linked, and adhere to solid surfaces well. The biggest advantage to this process is that polymers can be directly attached to a desired surface while the chains are growing, which reduces steps necessary for other coating processes such as grafting. This is very useful for pinhole-free coatings of 100 picometers to 1 micrometre thickness with solvent insoluble polymers.

In as early as the 1870s "polymers" formed by this process were known, but these polymers were initially thought of as undesirable byproducts associated with electric discharge, with little attention being given to their properties. It was not until the 1960s that the properties of these polymers where found to be useful. It was found that flawless thin polymeric coatings could be formed on metals, although for very thin films (<10mm) this has recently been shown to be an oversimplification. By selecting the monomer type and the energy density per monomer, known as the Yasuda parameter, the chemical composition and structure of the resulting thin film can be varied with a wide range. These films are usually inert, adhesive, and have low dielectric constants. Some common monomers polymerized by this method include styrene, ethylene, methacrylate and pyridine, just to name a few. The 1970s brought about many advances in plasma polymerization, including the polymerization of many different types of monomers. The mechanisms of deposition however were largely ignored until more recently. Since this time most attention devoted to plasma polymerization has been in the fields of coatings, but since it is difficult to control polymer structure, it has limited applications.

Basic Operating Mechanism

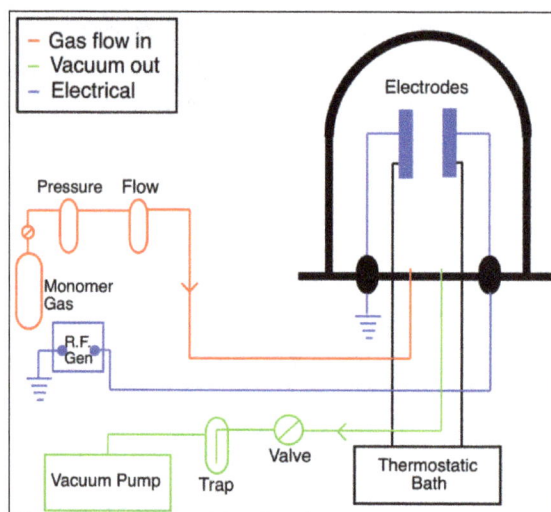

Schematic representation of basic internal electrode glow discharge polymerization apparatus.

Glow Discharge

Plasma consists of a mixture of electrons, ions, radicals, neutrals and photons. Some of these species are in local thermodynamic equilibrium, while others are not. Even for simple gases like argon this mixture can be complex. For plasmas of organic monomers, the complexity can rapidly increase as some components of the plasma fragment, while others interact and form larger species. Glow discharge is a technique in polymerization which forms free electrons which gain energy from an electric field, and then lose energy through collisions with neutral molecules in the gas phase. This leads to many chemically reactive species, which then lead to a plasma polymerization reaction. The electric discharge process for plasma polymerization is the "low-temperature plasma" method, because higher temperatures cause degradation. These plasmas are formed by a direct current, alternating current or radio frequency generator.

Types of Reactors

There are a few designs for apparatus used in plasma polymerization, one of which is the Bell (static type), in which monomer gas is put into the reaction chamber, but does not flow through the chamber. It comes in and polymerizes without removal. This type of reactor is shown in figure. This reactor has internal electrodes, and polymerization commonly takes place on the cathode side. All devices contain the thermostatic bath, which is used to regulate temperature, and a vacuum to regulate pressure.

Operation: The monomer gas comes into the Bell type reactor as a gaseous species, and then is put into the plasma state by the electrodes, in which the plasma may consist of radicals, anions and cations. These monomers are then polymerized on the cathode surface, or some other surface placed in the apparatus by different mechanisms. The deposited polymers then propagate off the surface and form growing chains with seemingly uniform consistency.

Another popular reactor type is the flow through reactor (continuous flow reactor), which also has internal electrodes, but this reactor allows monomer gas to flow through the reaction chamber as its name implies, which should give a more even coating for polymer film deposition. It has the advantage that more monomer keeps flowing into the reactor in order to deposit more polymer. It has the disadvantage of forming what is called "tail flame," which is when polymerization extends into the vacuum line.

A third popular type of reactor is the electrodeless. This uses an RF coil wrapped around the glass apparatus, which then uses a radio frequency generator to form the plasma inside of the housing without the use of direct electrodes. The polymer can then be deposited as it is pushed through this RF coil toward the vacuum end of the apparatus. This has the advantage of not having polymer building up on the electrode surface, which is desirable when polymerizing onto other surfaces.

A fourth type of system growing in popularity is the atmospheric-pressure plasma system, which is useful for depositing thin polymer films. This system bypasses the requirements for special hardware involving vacuums, which then makes it favorable for integrated industrial use. It has been shown that polymers formed at atmospheric-pressure can have similar properties for coatings as those found in the low-pressure systems.

Physical Process Characteristics

The formation of a plasma for polymerization depends on many of the following. An electron energy of 1–10 eV is required, with electron densities of 10^9 to 10^{12} per cubic centimeter, in order to form the desired plasma state. The formation of a low-temperature plasma is important; the electron temperatures are not equal to the gas temperatures and have a ratio of T_e/T_g of 10 to 100, so that this process can occur at near ambient temperatures, which is advantageous because polymers degrade at high temperatures, so if a high-temperature plasma was used the polymers would degrade after formation or would never be formed. This entails non-equilibrium plasmas, which means that charged monomer species have more kinetic energy than neutral monomer species, and cause the transfer of energy to a substrate instead of an uncharged monomer.

Kinetics

The kinetic rate of these reactions depends mostly on the monomer gas, which must be either gaseous or vaporized. However, other parameters are also important as well, such as power, pressure, flow rate, frequency, electrode gap and reactor configuration. Low flow rates usually only depend on the amount of reactive species present for polymerization, whereas high flow rates depend on the amount of time that is spent in the reactor. Therefore, the maximum rate of polymerization is somewhere in the middle.

The fastest reactions tend to be in the order of triple-bonded > double-bonded > single bonded molecules, and also lower molecular weight molecules are faster than higher ones. So acetylene is faster than ethylene, and ethylene is faster than propene, etc. The molecular weight factor in polymer deposition is dependent on the monomer flow rate, in which a higher molecular weight monomer typically near 200 g/mol needs a much higher flow rate of 15×10^4 g/cm^2, whereas lower molecular weights around 50 g/mol require a flow rate of only 5×10^4 g/cm^2. A heavy monomer therefore needs a faster flow, and would likely lead to increased pressures, decreasing polymerization rates.

Increased pressure tends to decrease polymerization rates reducing uniformity of deposition since uniformity is controlled by constant pressure. This is a reason that high-pressure plasma or atmospheric-pressure plasmas are not usually used in favor of low-pressure systems. At pressures greater than 1 torr, oligomers are formed on the electrode surface, and the monomers also on the surface can dissolve them to get a low degree of polymerization forming an oily substance. At low pressures, the reactive surfaces are low in monomer and facilitate growing high molecular weight polymers.

The rate of polymerization depends on input power, until power saturation occurs and the rate becomes independent of it. A narrower electrode gap also tends to increase polymerization rates because a higher electron density per unit area is formed. Polymerization rates also depend on the type of apparatus used for the process. In general, increasing the frequency of alternating current glow discharge up to about 5 kHz increases the rate due to the formation of more free radicals. After this frequency, inertial effects of colliding monomers inhibit polymerization. This forms the first plateau for polymerization frequencies. A second maximum in frequency occurs at 6 MHz, where side reactions are overcome again and the reaction occurs through free radicals diffused from plasma to the electrodes, at which point a second plateau is obtained. These parameters differ slightly for each monomer and must be optimized in-situ.

Synthetic Routes

Plasma contains many species such as ions, free radicals and electrons, so it is important to look at what contributes to the polymerization process most. The first suggested process by Westwood et al. was that of a cationic polymerization, since in a direct current system polymerization occurs mainly on the cathode. However, more investigation has led to the belief that the mechanism is more of a radical polymerization process, since radicals tend to be trapped in the films, and termination can be overcome by reinitiation of oligomers. Other kinetic studies also appear to support this theory.

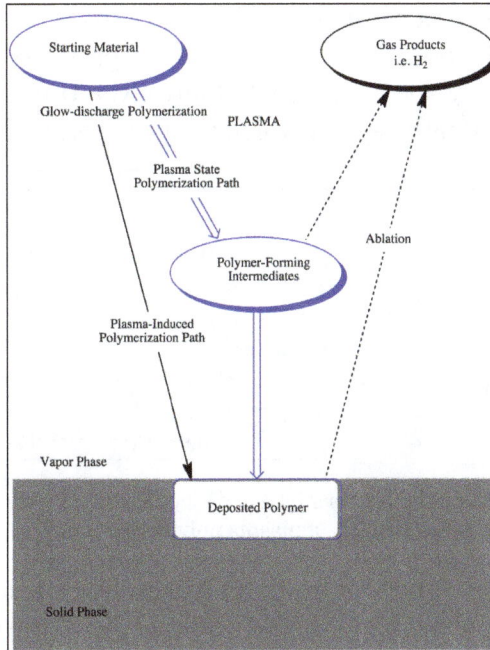

Schematic of plasma polymerization process possibilities,
w/blue representing dominant pathway.

However, since the mid 1990s a number of papers focusing on the formation of highly functionalized plasma polymers have postulated a more significant role for cations, particularly where the plasma sheath is collosionless. The assumption that the plasma ion density is low and consequently the ion flux to surfaces is low has been challenged, pointing out that ion flux is determined according to the Bohm sheath criterion i.e. ion flux is proportional to the square root of the electron temperature and not RT.

In polymerization, both gas phase and surface reactions occur, but mechanism differs between high and low frequencies. At high frequencies it occurs in reactive intermediates, whereas at low frequencies polymerization happens mainly on surfaces. As polymerization occurs, the pressure inside the chamber decreases in a closed system, since gas phase monomers go to solid polymers. An example diagram of the ways that polymerization can take place is shown in figure, wherein the most abundant pathway is shown in blue with double arrows, with side pathways shown in black. The ablation occurs by gas formation during polymerization. Polymerization has two pathways, either the plasma state or plasma induced processes, which both lead to deposited polymer.

Polymers can be deposited on many substrates other than the electrode surfaces, such as glass, other organic polymers or metals, when either a surface is placed in front of the electrodes, or placed in the middle between them. The ability for them to build off of electrode surfaces is likely to be an electrostatic interaction, while on other surfaces covalent attachment is possible.

Polymerization is likely to take place through either ionic and/or radical processes which are initiated by plasma formed from the glow discharge. The classic view presented by Yasuda based upon thermal initiation of parylene polymerization is that there are many propagating species present at any given time as shown in figure. This figure shows two different pathways by which the polymerization may take place.

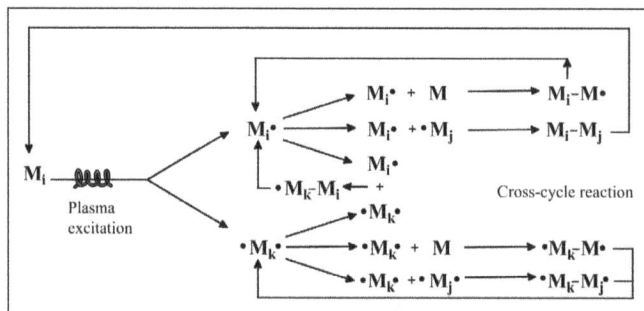

Schematic representation of bicyclic step-growth
mechanism of plasma polymerization.

The first pathway is a monofunctionalization process, bears resemblance to a standard free radical polymerization mechanism $(M\cdot)$- although with the caveat that the reactive species may be ionic and not necessarily radical. The second pathway refers to a difunctional mechanism, which by example may contain a cationic and a radical propagating center on the same monomer $(\cdot M\cdot)$. A consequence is that 'polymer' can grow in multiple directions by multiple pathways off one species, such as a surface or other monomer. This possibility let Yasuda to term the mechanism as a very rapid step-growth polymerization. In the diagram, M_x refers to the original monomer molecule or any of many dissociation products such as chlorine, fluorine and hydrogen. The $M\cdot$ species refers to those that are activated and capable of participating in reactions to form new covalent bonds. The $\cdot M\cdot$ species refers to an activated difunctional monomer species. The subscripts i, j, and k show the sizes of the different species involved. Even though radicals represent the activated species, any ion or radical could be used in the polymerization. As can be seen here, plasma polymerization is a very complex process, with many parameters effecting everything from rate to chain length.

Selection, or the favouring of one particular pathway can be achieved by altering the plasma parameters. For example, pulsed plasma with selected monomers appears to favour much more regular polymer structures and it has been postulated these grow by a mechanism akin to (radical) chain growth in the plasma off-time.

Common Monomers/Polymers

Common monomers	
Name	Structure
Thiophene	

1,7-Octadiene	
Pyridine	
Acrylonitrile	
Furan	
Styrene	
Acetylene	$H-C\equiv C-H$
2-Methyloxazoline	
Tetramethyldisiloxane	

Monomers

As can be seen in the monomer table, many simple monomers are readily polymerized by this method, but most must be smaller ionizable species because they have to be able to go into the plasma state. Though monomers with multiple bonds polymerize readily, it is not a necessary requirement, as ethane, silicones and many others polymerize also. There are also other stipulations that exist. Yasuda et al. studied 28 monomers and found that those containing aromatic groups, silicon, olefinic group or nitrogen (NH, NH_2, CN) were readily polymerizable, while those containing oxygen, halides, aliphatic hydrocarbons and cyclic hydrocarbons where decomposed more readily. The latter compounds have more ablation or side reactions present, which inhibit stable polymer formation. It is also possible to incorporate N_2, H_2O, and CO into copolymers of styrene.

Plasma polymers can be thought of as a type of graft polymers since they are grown off of a substrate. These polymers are known to form nearly uniform surface deposition, which is one of their desirable properties. Polymers formed from this process often cross-link and form branches due to the multiple propagating species present in the plasma. This often leads to very insoluble polymers, which gives an advantage to this process, since hyperbranched polymers can be deposited directly without solvent.

Polymers

Common polymers include: polythiophene, polyhexafluoropropylene, polytetramethyltin, polyhexamethyldisiloxane, polytetramethyldisiloxane, polypyridine, polyfuran, and poly-2-methyloxazoline.

The following are listed in order of decreasing rate of polymerization: polystyrene, polymethyl styrene, polycyclopentadiene, polyacrylate, polyethyl acrylate, polymethyl methacrylate, polyvinyl acetate, polyisoprene, polyisobutene, and polyethylene.

Nearly all polymers created by this method have excellent appearance, are clear, and are significantly cross-linked. Linear polymers are not formed readily by plasma polymerization methods based on propagating species. Many other polymers could be formed by this method.

General Characteristics of Plasma Polymers

The properties of plasma polymers differ greatly from those of conventional polymers. While both types are dependent on the chemical properties of the monomer, the properties of plasma polymers depend more greatly on the design of the reactor and the chemical and physical characteristics of the substrate on which the plasma polymer is deposited. The location within the reactor where the deposition occurs also has an effect on the resultant polymer's properties. In fact by using plasma polymerization with a single monomer and varying the reactor, substrate, etc. a variety of polymers, each having different physical and chemical properties, can be prepared. The large dependence of the polymer features on these factors make it difficult to assign a set of basic characteristics, but a few common properties that set plasma polymers apart from conventional polymers do exist.

The most significant difference between conventional polymers and plasma polymers is that plasma polymers do not contain regular repeating units. Due to the number of different propagating species present at any one time, the resultant polymer chains are highly branched and are randomly terminated with a high degree of cross-linking. An example of a proposed structure for plasma polymerized ethylene demonstrating a large extend of cross-linking and branching is shown in Figure.

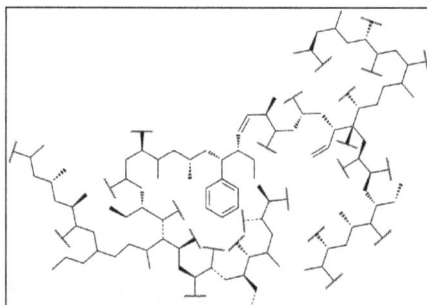

Hypothesized model of plasma-polymerized ethylene film.

All plasma polymers contain free radicals as well. The amount of free radicals present varies between polymers and is dependent on the chemical structure of the monomer. Because the formation of the trapped free radicals is tied to the growth mechanism of the plasma polymers, the overall properties of the polymers directly correlate to the number of free radicals.

Plasma polymers also contain an internal stress. If a thick layer (e.g. 1 μm) of a plasma polymer is deposited on a glass slide, the plasma polymer will buckle and frequently crack. The curling is attributed to an internal stress formed in the plasma polymer during the polymer deposition. The degree of curling is dependent on the monomer as well as the conditions of the plasma polymerization.

Most plasma polymers are insoluble and infusible. These properties are due to the large amount of cross-linking in the polymers. Consequently, the kinetic path length for these polymers must be sufficiently long, so these properties can be controlled to a point.

The permeabilities of plasma polymers also differ greatly from those of conventional polymers. Because of the absence of large-scale segmental mobility and the high degree of cross-linking within the polymers, the permeation of small molecules does not strictly follow the typical mechanisms of "solution-diffusion" or molecular-level sieve for such small permeants. Really the permeability characteristics of plasma polymers falls between these two ideal cases.

A final common characteristic of plasma polymers is the adhesion ability. The specifics of the adhesion ability for a given plasma polymer, such as thickness and characteristics of the surface layer, again are particular for a given plasma polymer and few generalizations can be made.

Advantages and Disadvantages

Plasma polymerization offers a number of advantages over other polymerization methods and in general. The most significant advantage of plasma polymerization is its ability to produce polymer films of organic compounds that do not polymerize under normal chemical polymerization conditions. Nearly all monomers, even saturated hydrocarbons and organic compounds without a polymerizable structure such as a double bond, can be polymerized with this technique.

A second advantage is the ease of application of the polymers as coatings versus conventional coating processes. While coating a substrate with conventional polymers requires a number of steps, plasma polymerization accomplishes all these in essentially a single step. This leads to a cleaner and 'greener' synthesis and coating process, since no solvent is needed during the polymer preparation and no cleaning of the resultant polymer is needed either. Another 'green' aspect of the synthesis is that no initiator is needed for the polymer preparation since reusable electrodes cause the reaction to proceed. The resultant polymer coatings also have a number of advantages over typical coatings. These advantages include being nearly pinhole free, highly dense, and that the thickness of the coating can easily be varied.

There are also a number of disadvantages relating to plasma polymerization versus conventional methods. The most significant disadvantage is the high cost of the process. A vacuum system is required for the polymerization, significantly increasing the set up price.

Another disadvantage is due to the complexity of plasma processes. Because of the complexity it is not easy to achieve a good control over the chemical composition of the surface after modification. The influence of process parameters on the chemical composition of the resultant polymer means it can take a long time to determine the optimal conditions. The complexity of the process also makes it impossible to theorize what the resultant polymer will look like, unlike conventional polymers which can be easily determined based on the monomer.

Applications

The advantages offered by plasma polymerization have resulted in substantial research on the applications of these polymers. The vastly different chemical and mechanical properties offered by polymers formed with plasma polymerization means they can be applied to countless different systems. Applications ranging from adhesion, composite materials, protective coatings, printing, membranes, biomedical applications, water purification and so on have all been studied.

Of particular interest since the 1980s has been the deposition of functionalized plasma polymer films. For example, functionalized films are used as a means of improving biocompatibility for biological implants6 and for producing super-hydrophobic coatings. They have also been extensively employed in biomaterials for cell attachment, protein binding and as anti-fouling surfaces. Through the use of low power and pressure plasma, high functional retention can be achieved which has led to substantial improvements in the biocompatibility of some products, a simple example being the development of extended wear contact lenses. Due to these successes, the huge potential of functional plasma polymers is slowly being realised by workers in previously unrelated fields such as water treatment and wound management.Emerging technologies such as nanopatterning, 3D scaffolds, micro-channel coating and microencapsulation are now also utilizing functionalized plasma polymers, areas for which traditional polymers are often unsuitable

A significant area of research has been on the use of plasma polymer films as permeation membranes. The permeability characteristics of plasma polymers deposited on porous substrates are different than usual polymer films. The characteristics depend on the deposition and polymerization mechanism. Plasma polymers as membranes for separation of oxygen and nitrogen, ethanol and water, and water vapor permeation have all been studied. The application of plasma polymerized thin films as reverse osmosis membranes has received considerable attention as well. Yasuda et al. have shown membranes prepared with plasma polymerization made from nitrogen containing monomers can yield up to 98% salt rejection with a flux of 6.4 gallons/ft^2 a day. Further research has shown that varying the monomers of the membrane offer other properties as well, such as chlorine resistance.

Plasma-polymerized films have also found electrical applications. Given that plasma polymers frequently contain many polar groups, which form when the radicals react with oxygen in air during the polymerization process, the plasma polymers were expected to be good dielectric materials in thin film form. Studies have shown that the plasma polymers generally do in fact have a higher dielectric property. Some plasma polymers have been applied as chemical sensory devices due to their electrical properties. Plasma polymers have been studied as chemical sensory devices for humidity, propane, and carbon dioxide amongst others. Thus far issues with instability against aging and humidity have limited their commercial applications.

The application of plasma polymers as coatings has also been studied. Plasma polymers formed from tetramethoxysilane have been studied as protective coatings and have shown to increase the hardness of polyethylene and polycarbonate. The use of plasma polymers to coat plastic lenses is increasing in popularity. Plasma depositions are able to easily coat curved materials with a good uniformity, such as those of bifocals. The different plasma polymers used can be not only scratch resistant, but also hydrophobic leading to anti-fogging effects. Plasma polymer surfaces with tunable wettability and reversibly switchable pH-responsiveness have shown the promising prospects due to their unique property in applications, such as drug delivery, biomaterial engineering, oil/water separation processes, sensors, and biofuel cells.

PLASMA ELECTROLYTIC OXIDATION

Plasma electrolytic oxidation (PEO), also known as micro-arc oxidation (MAO), is a bath-based method of producing ceramic layers on the surface of light alloys. PEO surface coatings are characterised by their wear resistance, corrosion resistance and thermal and chemical stability. The method is suitable for alloys of high aluminium, magnesium and titanium composition, but can also be applied to other metals such as zirconium, tantalum, niobium, hafnium and cobalt.

Electrolytic oxidation without the use of plasma - anodising - has been a prevalent technique over many years. The introduction of plasma fundamentally alters alters coating and performance characteristics in stressful end-use applications.

The use of plasma introduces several benefits:

- Development of harder ceramic phases (including crystallisation).

- Chemical passivity - most PEO ceramics are chemically inert.

- Incorporation of elements from the electrolyte into the ceramic to give different properties.

- Reduced stiffness gives high adhesion under mechanical strain or thermal cycling.

- Crack-free edges.

Surface coatings formed through plasma electrolytic oxidation can offer 2-4x more hardness than hard anodizing or steel, and provide increased wear resistance. The combination of these qualities and the comprehensiveness of protection has made PEO a breakthrough surface engineering innovation.

Plasma Electrolytic Oxidation Process

The process generally follows these three stages:

Oxidation of the Substrate

As occurs during anodising, a component is submerged in a bath of electrolyte. Bath compositions differ based on the desired characteristics of the PEO coating, but is usually a proprietary dilute aqueous solution. This is free from chrome and other heavy metals. Additionally, the solution composition is disposable and clean, in contrast to hard anodising techniques which employ sulphuric acid (H_2SO_4).

Depending on the desired coating characteristics, different electrical regimes can be employed. For example, alternating the polarity of an aluminium substrate can achieve variations of growth formation. Higher voltages are typically used to create plasma discharge.

Plasma Modification

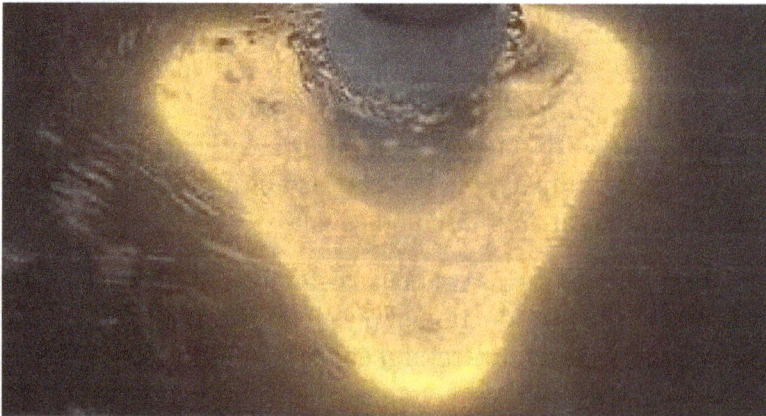

Plasma discharge around a component
immersed in electrolyte.

In conventional anodising, the coating growth mechanism causes through-thickness cracks or fissures in the protective layer on corners or uneven surfaces. Additional seals or treatments are necessary to increase the corrosion resistance capabilities of hard anodised components for this reason. These also reduce the fatigue strength of a component, acting as stress raisers.

With PEO, plasma is used to modify the coating during the growth process. This alters the microstructure, resulting in no through-thickness cracks, and provide consequent benefits in corrosion resistance and fatigue strength.

In many cases, the plasma also causes crystallisation of the oxide layer, increasing hardness and potential for wear resistance. Micro and nanocrystals such as Al_2O_2 corundum in aluminium, periclase on magnesium and anatase/rutile in titanium can be introduced into the ceramic layer for enhanced hardness.

Plasma modification also creates other attractive features such as chemical passivity, low stiffness and thermal stability.

Incorporation of Electrolyte Elements into the Coating

The plasma modification process enables elements of the electrolyte to be incorporated within the ceramic composition of the layer. This means engineers can select elements to tailor coating properties specifically for the application of the component. For example, black or white coatings can be created this way, as can basic properties such as porosity, hardness and process efficiency.

It is this level of flexibility and controllability over the coating characteristics that have made PEO such an attractive choice for engineers.

Process Parameters

The processes involved in PEO are highly flexible, particularly when compared to alternatives such as hard anodising. This opens up a wide range of potential surface coating properties, which can be adapted and tailored to best suit the end use application of a component:

- Pre-treatment is simpler for PEO. In many cases, aluminium components may only require a light degrease prior to treatment. By comparison, most anodising and plating processes will need a cleaner surface, necessitating degreasing, etching and desmut steps to ensure a high quality coating.

- Applied electrical parameters can be adjusted based on a preferred surface morphology. Coating morphologies can be created with mean length scales from nanometer to micron range.

- Process chemistry can be developed to create different coatings and treat different metals and alloys.

When optimising a PEO layer for specific use, the ability to enhance certain coating characteristics is vital in the overall quality and sustainability of a component.

Coating Characteristics

PEO's unique and flexible process produces highly protective layers that can be enhanced for performance in specific usage. With the added bonus of being a clean process, free of heavy metals, it's unsurprising that widespread interest has been sparked for the technique.

Coverage

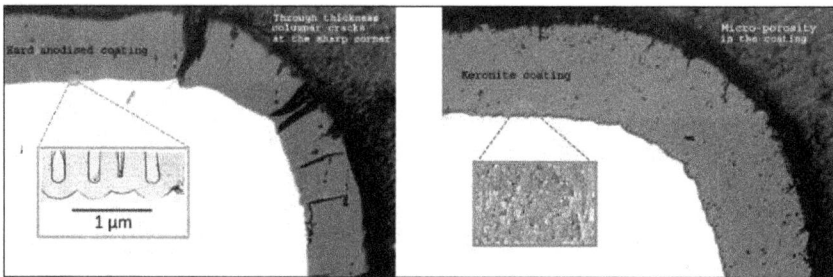

Comparative layer structures of Hard Anodising Coatings (left) and PEO (right).

In general, PEO coatings are characterised by good coverage of the component compared to line-of-sight processes such as painting, powder coating and plasma or flame spray techniques. Also, the insulating properties of the coating ensure good uniformity at corners and edges. Paints tend to thin at corners, whereas electroplating techniques tend to thicken at corners.

As illustrated in the SEM images in figure, PEO layers are characterised by their complex microstructure. The presence of irregularly shaped microcrystals and other features provide more comprehensive protection on corners than through-thickness cracks clearly visible in hard anodised coatings.

Extreme Hardness

As light alloys become more desirable to work with, both in terms of performance and cost, the need for innovative surface coatings has increased.

One such quality is the extreme hardness created in PEO. Typical coatings on aluminium are harder than steel (1600HV vs 500HV), yet the component itself could be up to 66% lighter. The performance enhancing characteristics of PEO coatings have enabled light alloys - even magnesium - to prominently feature in aerospace and automotive applications.

This extreme hardness is gained through a combination of crystallisation (of the oxides) and co-deposition of elements from the electrolyte in the ceramic layer. Aluminium, for example, can generate α-Al_2O_3 crystalline phases on AA7075, with hardness up to 2000HV, which outperforms steels on pin-on-disc tests.

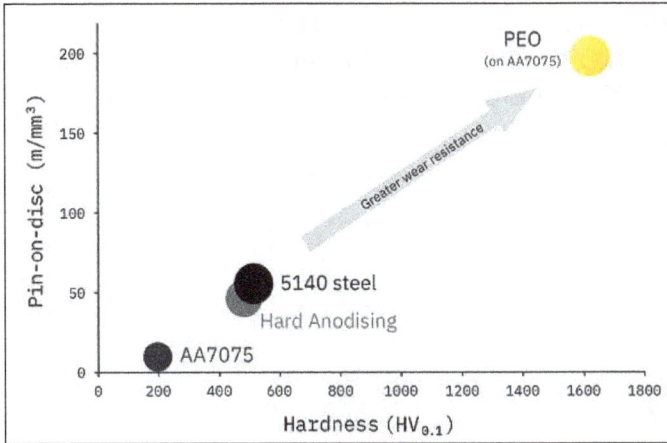

Hardness of PEO layers in pin-on-disc tests.

Corrosion Resistance

For optimal corrosion resistance performance, PEO works best as a pretreatment for subsequent sealers, paints and other polymers.

Polyester Powder Coat on a PEO coating
on a Magnesium substrate.

In figure illustrates a polyester powder coat image applied to a PEO layer on a magnesium alloy substrate. The strong bond between the two layers enhances corrosion resistance capabilities, also forming scratch resistant qualities. The polyester coating effectively fills the pore architecture created during the formation of the PEO layer.

Generally, PEO is good for different types of bonding because its reticulated microstructure creates a physical 'key' and does not rely upon chemical compatibility between additional coats, unlike alternative coating mechanisms. The same principle applies to adhesion of oil and other lubricants in sliding wear applications.

Blanchard et al., in a study of 25 different surface treatments for Magnesium in terms of corrosion, found PEO to surpass the Chromium VI and Chromium-free alternatives tested. The test results are illustrated in figure , with F, G, H, I and J being derivatives of plasma electrolytic oxidation.

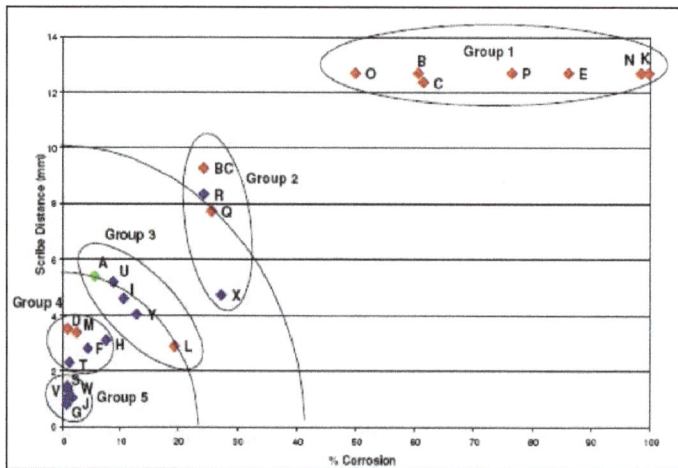

Evaluation of Corrosion Protection Methods for Magnesium Alloys in Automotive.

High Strain Tolerance

High hardness alone does not necessarily provide comprehensive wear resistance capabilities. Compliance is also a very valuable property, allowing for some deformation of the component under deflection or thermal expansion without placing undue stress into the coating-metal surface.

Hardness and appropriate levels of compliance combined increase the wear-resistance of a substrate. Again, PEO's unique microstructure gives the material high fracture toughness, reducing the potential for cracking under force, which means ceramic surface coatings provide excellent performance in tribology applications.

Environmental Friendliness

Coating and surface treatments that employ heavy metals (such as nickel, cobalt and frequently chromium) often involve high-hazard chemicals. Regulations surrounding the use of these are tightening, thus making them more difficult to work with. Conventional anodising employ strong acids, raising safety issues in use, transportation and disposal.

PEO is an environmentally safe option. Electrolytic baths are typically low concentration, chemically benign, aqueous solutions. Process waste streams can typically be discharged directly to drain after pH adjustment, so operating licenses are easily obtained. It is for this reason the technique has become popular in highly technical industries that require high performance surface coats in challenging environments.

PLASMA CUTTING

Plasma cutting is a non-conventional method which can be used in good conditions to cut different materials with a plasma torch. In this process, an inert gas is blown with high speed out of a nozzle; at the same time, an electrical arc is formed through that gas from the nozzle to the surface, being cut turning some of that gas to the plasma. The plasma melts the material being cut and swiftly moves blowing molten metal away from the cut. A summary analysis of composite materials cutting methods, based on the specialty literature, was made.

This work presents some results regarding quality surface, surface integrity and heat affected zone of one laminar composite material (sandwich panels). This material is composed of two aluminium plates (0.3 mm) with a polyethylene core.

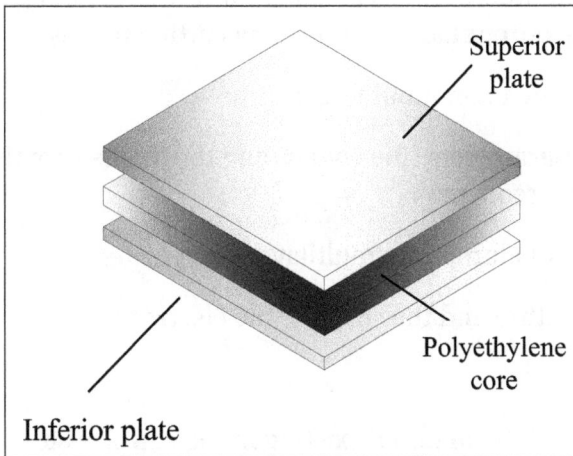

Composite material used for research.

Differences between Traditional Methods and Plasma Cutting

Classical Methods

Because machining of composite materials imposes special demands on the geometry and materials used for cutting tools, as well as special requirements regarding shape complexity which must be obtained, these methods cannot be used in all the situations. Usually, in case of aluminium-polyethylene sandwich panels, which is not a very rough material, traditional methods, based on shear deformation, can be used. But, when one has to obtain very complex shapes from large panels, these methods can no longer be used. One advantage of classical methods is that the machined area is not thermally affected. Another aspect is high tool wear, in case of hard multilayer composites. Regarding the cost of traditional methods, it is clear that for a small production and for simple shapes, it is more economical to use them.

Plasma Cutting Method

Because plasma cutters produce a very hot and localized jet to cut with, they are very useful to cut sheet metal plates in curved or angled shapes. Plasma cutting machines are used in CNC machinery. In this way, the entire process can be controlled and optimized by computer, obtaining clean sharp cuts. By the use of multi-axis CNC machines, plasma equipment is able not only to cut complex shapes, but welding of different materials can also be performed. Combination of CNC technology with smaller nozzles for a thinner plasma arc allows for parts that require no finishing operation to be obtained.

Plasma Cutting – Systemic Approach

For a good appreciation of plasma cutting conditions, one has to know which the factors that can affect the process are. The diagram presented in figure shows the input, the disturbing and the output factors for plasma cutting process.

As it can be observed, the main input factors are:

- The electrical parameters (the voltage and the intensity of the electrical current, the oscillations frequency).

- The properties of aluminium multilayer composite.

- The material nature and the shape of the electrode sharp end.

- The gap size.

- The properties of the material existing in the gap size etc.

The factors able to disturb the plasma cutting process can be considered: the work piece material non-homogeneity, the current intensity variation, the atmospheric conditions from the working zone etc.

The output parameters can be considered:

- The cutting speed.

- The width of the cut channel.

- The thickness of heat affected area.

- The surface quality and integrity etc.

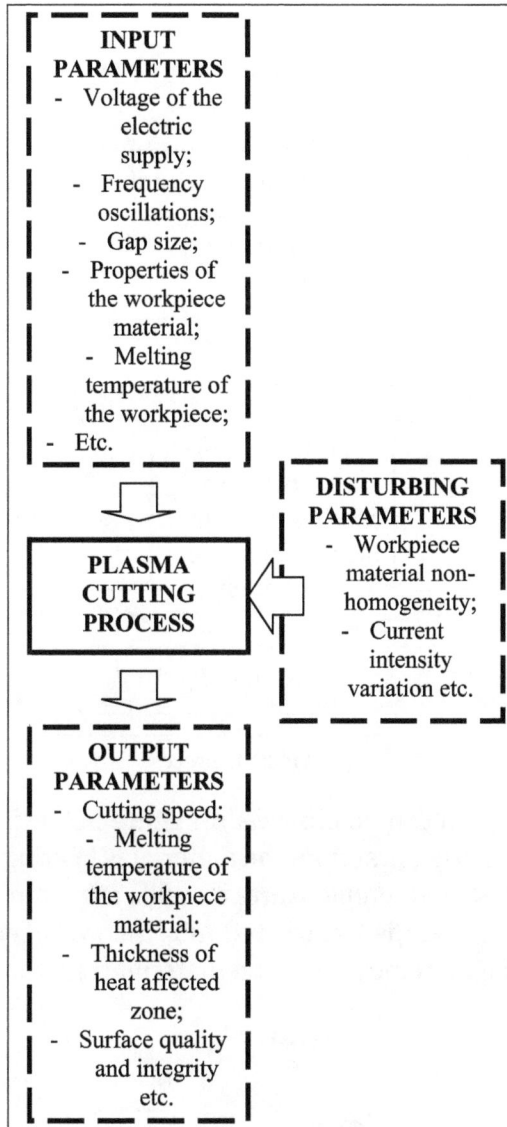

Factors able to affect the plasma cutting process.

Physical Modeling

Plasma cutting of the workpiece is the result of melting/vaporizing of material through a very hot cylindrical (theoretical) plasma beam which burns and melts through the material. One of the most important problems occurring as result of heat transfer from plasma column to the workpiece is the deformation of the cut edges after the material is cut and then cooled. During the cut operation, many physical phenomena occur in this process: heat conduction, convection, radiation effects, mechanical deformation, phase transition etc. The principle of plasma cutting means to focus a lot of power on a small area of workpiece surface, in this way producing an intense surface heating.

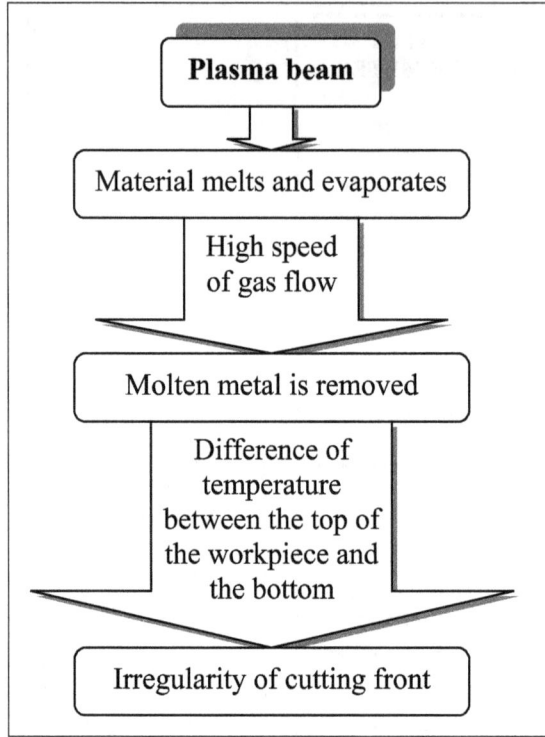

A typical plasma process.

The first phenomenon which can be observed in figure below is the absorption of energy by the workpiece material, first time the material is heating. At a certain moment, when the material reaches melt temperature, it melts, afterwards resulting a solid-liquid phase. Another part of heat is transferred into the workpiece by conduction. The high velocity of plasma beam removes molten material from the top of the cut.

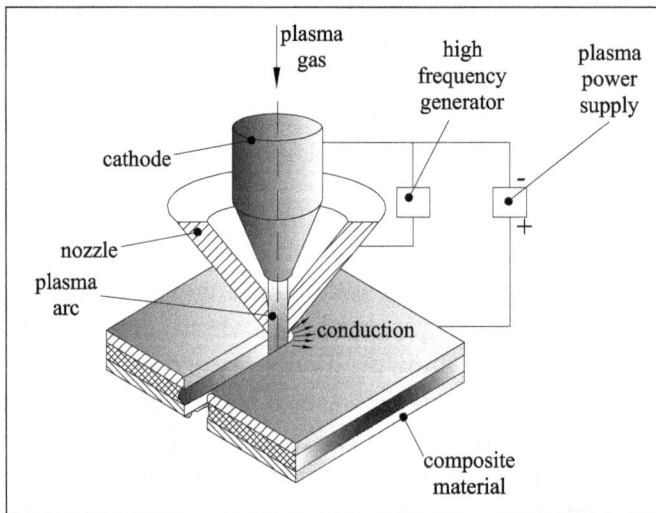

Thermal cutting.

Experimental Procedure

The experiment was performed in a non-conventional laboratory using an Italian plasma arc machine Telwin PL 36/2. The plasma cutting power source characteristics are presented in table below.

Table: Plasma cutting power source characteristics.

INPUT	Main voltage 230 (V)	Absorbed current 23 (A)	Power 7 (kVA)
OUTPUT	Rated cutting voltage 95 (V)	Rated cutting current 30 (A)	Cutting capacity for steel 0,6-0,3 (mm)

Some preliminary cuts were made in order to fix the measurements procedure and to establish the optimal conditions. The main problem is how the polyethylene core will be influenced and what zone will be melted near the cut edge. The optimal parameters used, which can be adjusted, are: pressure – 0.41 MPa; feed – 50 m/min; distance between workpiece and nozzle – 5 mm. A visual inspection of the plate confirms that when instability of plasma arc occurs, the lack of energy contribution from the plasma jet leads to irregular melting and removal of material. This phenomenon occurs when the torch stand is too high (more than 5 mm in this case) and the arc cannot reach the bottom of the workpiece with required energy density and constriction.

To improve the productivity and to obtain different shapes with complex angles and geometries, CNC plasma cutting machinery can be used. In this study a sandwich composite material was used. The main problem in cutting this material is the polyethylene core. Because of its low melting point, this core is melted and removed near the cut edge. In case of industrial high power plasma cutting machine, if the cutting speed is bigger, the polyethylene core is less thermally influenced. In figure below the heat flux density and quality of cut can be observed.

The kerf dimension is approximately 2 mm. The thermally influenced area of polyethylene core is around 5 mm and the dimension of the burr in the bottom side of the workpiece is around 1 mm. The plasma arc cutting technology can be used in case of aluminium composite plates with polyethylene core, but the quality of the core is not very good. To improve quality of the cut, high power plasma machinery which can perform on a high cutting speed must be used. The cutting speed showed a significant effect. Using a high cutting speed, the plasma arc will spend less time on workpiece surface and the polyethylene core will be less thermally affected. This method is usually used in industrial applications for big panels and when other methods cannot be applied. In order to optimize and to obtain a very good quality for all sides (top side, bottom side and core), it is necessary to vary the cutting speed and the arc voltage.

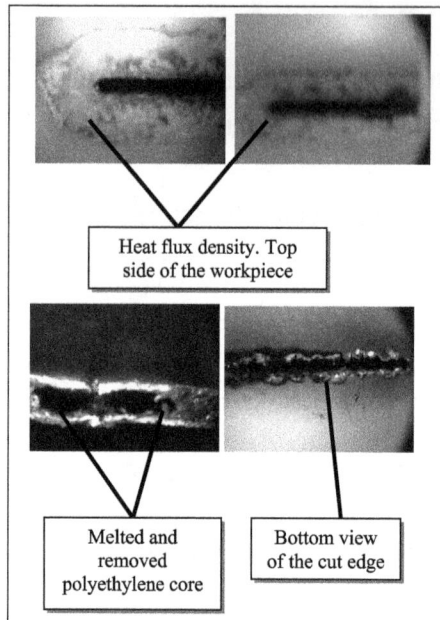

Examples of aluminium sandwich composite panels
cut with plasma arc machine.

Another non-conventional method, which can be used with good results for cutting complex shapes, is laser cutting and water jet technology. In case of this material, a good way to cut it by the use of CNC technology seems to be water jet machinery. In this way, the polyethylene core will no longer be thermally affected. Therefore, future investigation can be focused on high power plasma machine cutting, from industrial area, laser cutting and water jet technology.

PLASMA GASIFICATION

Plasma gasification or plasma-assisted gasification can be used to convert carbon-containing materials to synthesis gas to generate power and other useful products, such as transportation fuels. In an effort to reduce both the economic and environmental costs of managing municipal solid waste, (which includes construction and demolition wastes) a number of cities are working with plasma gasification companies to send their wastes to these facilities. One city in Japan gasifies its wastes to produce power. In addition, various industries that generate hazardous wastes as part of their manufacturing processes (such as the chemical and refining industries) are examining plasma gasification as a cost-effective means of managing those wastes streams.

Plasma is an ionized gas that is formed when an electrical discharge passes through a gas. The resultant flash from lightning is an example of plasma found in nature. Plasma torches and arcs convert electrical energy into intense thermal (heat) energy. Plasma torches and arcs can generate temperatures up to 10,000 degrees Fahrenheit. When used in a gasification plant, plasma torches and arcs generate this intense heat, which initiates and supplements the gasification reactions, and can even increase the rate of those reactions, making gasification more efficient.

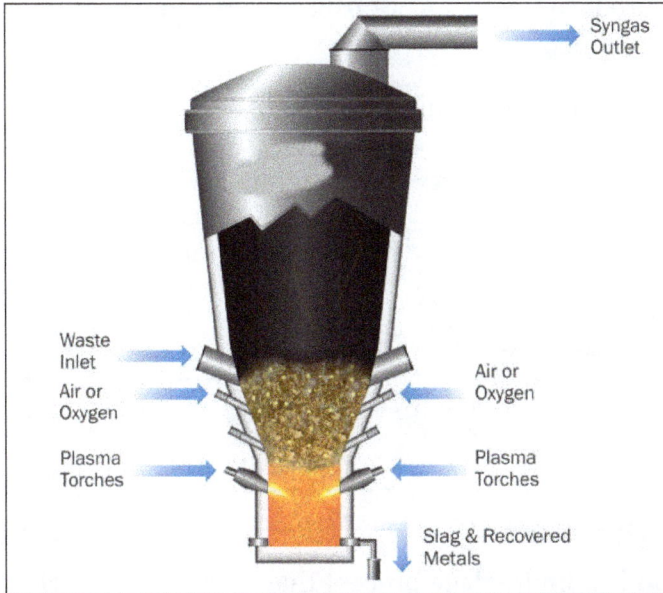

Inside the gasifier, the hot gases from the plasma torch or arc contact the feedstock, such as municipal solid waste, auto shredder wastes, medical waste, biomass or hazardous waste, heating it to more than 3,000 degrees Fahrenheit. This extreme heat maintains the gasification reactions, which break apart the chemical bonds of the feedstock and converts them to a syngas. The syngas consists primarily of carbon monoxide and hydrogen—the basic building blocks for chemicals, fertilizers, substitute natural gas, and liquid transportation fuels. The syngas can also be sent to gas turbines or reciprocating engines to produce electricity or combusted to produce steam for a steam turbine-generator.

Because the feedstocks reacting within the gasifier are converted into their basic elements, even hazardous waste becomes a useful syngas. Inorganic materials in the feedstock are melted and fused into a glassy-like slag, which is nonhazardous and can be used in a variety of applications, such as roadbed construction and roofing materials.

Key benefits of plasma gasification:

- It unlocks the greatest amount of energy from waste.

- Feedstocks can be mixed, such as municipal solid waste, biomass, tires, hazardous waste, and auto shredder waste.

- It does not generate methane, a potent greenhouse gas.

- It is not incineration and therefore doesn't produce leachable bottom ash or fly ash.

- It reduces the need for landfilling of waste.

- It produces syngas, which can be combusted in a gas turbine or reciprocating engines to produce electricity or further processed into chemicals, fertilizers, or transportation fuels.

- It has low environmental emissions.

Plasma Gasification Process

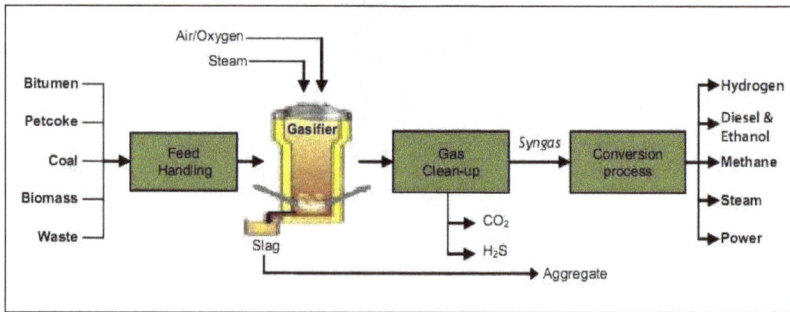

Plasma gasification is a multi-stage process that starts with a variety of inputs ranging from garbage to coal to plant matter, but can include any hazardous waste or carbon-based material. The first step is to process the feed stock to make it more uniform and dry for gasification; in the case of MSW, it will be shredded and valuable recyclables sorted out. The second step is gasification, where extreme heat from the plasma torches is applied inside a sealed, air-controlled reactor. During gasification, carbon-based materials break down into gasses. The extreme heat from the torches causes all the inorganic materials to melt and form slag. The extreme heat also causes all the hazards and poisons to be completely destroyed. The technology has its roots in hazardous-waste destruction. The third stage is the gas cleanup and heat recovery. The gasses are scrubbed of all impurities, forming a very clean fuel gas. Heat exchangers are used to recycle the heat back into the system in the form of steam and electricity. The final stage is fuel production which can range from electricity to liquid fuels like ethanol, hydrogen, natural gas, or chemicals and polymers.

Gasification is an old industrial process that uses heat in an oxygen-starved and pressurized environment to break down carbon-based materials into fuel gasses. There is a huge variety of gasification equipment and techniques that are tailored to deal with a wide variety of raw materials. Any material made from carbon is suitable for gasification, and the most common materials used are coal and biomass, such as wood or agricultural wastes. Coal gasification is a major industry with a long history in use to produce fuels ranging from old-fashioned "town-gas" to ultra-clean diesel and chemi-cals. Modern clean-coal plants all use gasification systems. Many pathways for producing cellulosic ethanol utilize biomass gasification to break down wood waste and other non-food crops to make a gas that can be processed into ethanol. Gasification is more environmentally sound and more fuel-efficient than typical combustion systems, and is being heavily promoted by the energy industry as an environmentally sound means to utilize coal and other unconventional hydrocarbons such as tar sands. As a means to treat garbage, gasification is far superior to incineration, both environmentally and in net energy production.

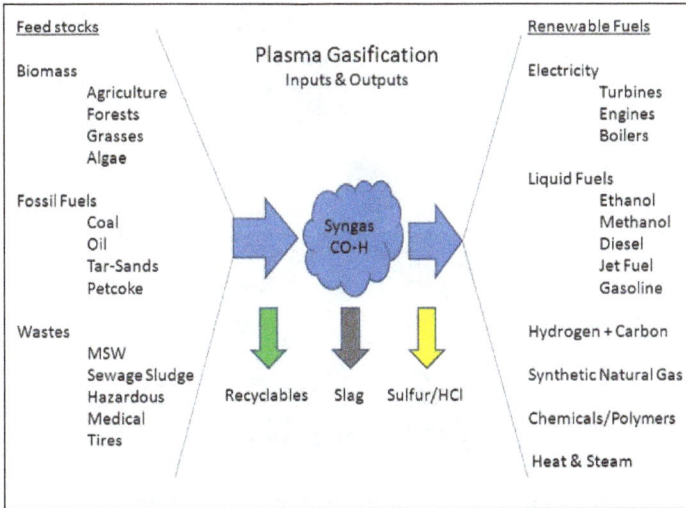

Plasma Gasification
Inputs & Outputs

Feed stocks

Biomass
  Agriculture
  Forests
  Grasses
  Algae

Fossil Fuels
  Coal
  Oil
  Tar-Sands
  Petcoke

Wastes
  MSW
  Sewage Sludge
  Hazardous
  Medical
  Tires

Syngas CO-H

Recyclables   Slag   Sulfur/HCl

Renewable Fuels

Electricity
  Turbines
  Engines
  Boilers

Liquid Fuels
  Ethanol
  Methanol
  Diesel
  Jet Fuel
  Gasoline

Hydrogen + Carbon

Synthetic Natural Gas

Chemicals/Polymers

Heat & Steam

Gasification is closely related to combustion and pyrolysis, but there are important distinctions. Gasification is like starved-air burning because oxygen is strictly controlled and limited so that as heat is applied the feedstock is not allowed to actually burn. Instead of combusting, the raw materials break down and go through the process of pyrolysis that produces char and tar. At its simplest form, pyrolysis is commonly used to produce charcoal from wood. As the process continues and the heat is taken higher, the char and tar completely break down into gasses. Depending on the process used and the precise chemistry, the resulting gas may come in a few different forms: synthesis gas, producer gas, town gas, wood gas or others.

Synthesis gas, also known as syngas, will be the focus of this report. Syngas is a simple blend of CO-H, carbon monoxide and hydrogen. This gas burns very cleanly with properties very similar to natural gas, although with less heating value. Syngas can be burned to produce heat and steam, or electricity through the use of boilers, engines, and turbines. Alternatively, syngas can be processed using catalysts and refined into a variety of liquid fuels. Fischer-Tropsch synthesis was invented in the 1920's and has been used heavily since WWII to produce gasoline and diesel from coal. Traditionally, coal-to-liquids has been much more expensive than petroleum, but the recent rise in oil prices has made many unconventional energy technologies cost-competitive and new catalysts are being developed to economically produce ethanol. Syngas can also be used to produce hydrogen and is considered a primary pathway to a possible hydrogen economy by the U.S. Dept. of Energy. Syngas can be upgraded into synthetic natural gas or used to make many different industrial chemicals.

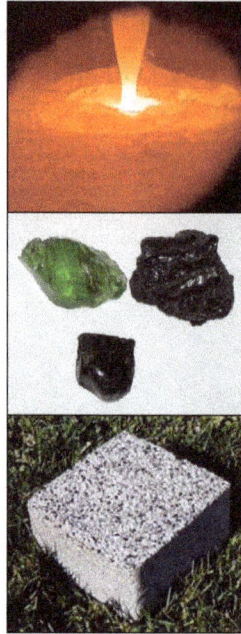

Plasma gasification refers to a range of techniques that utilize plasma torches or plasma arcs to generate extreme temperatures that are particularly effective for highly efficient gasification. Plasma is a superheated column of electrically conductive gas. In nature, plasma is found in lightning and on the surface of the sun. Plasma torches burn at temperatures approaching 10,000 °F and can reliably destroy any materials found here on earth with the exception of nuclear waste, since radioactive isotopes are not broken down by heat. Plasma torches are typically used in foundries to melt and cut metals, and similar electric-powered furnaces melt metals by the ton. When utilized for waste treatment, plasma torches are very efficient at causing organic and carbonaceous materials to vaporize into gas. Non-organic materials are melted and cool into a vitrified glass. Waste gasification typically operates at temperatures of 1500 °C and at those temperatures materials are subject to a process called molecular disassociation, which means that their molecular bonds are broken down, and in the process all toxins and organic poisons are destroyed. Plasma torches have been used for many years to destroy chemical weapons and toxic wastes like PCBs and asbestos, but it is only recently that these processes have been optimized for energy capture and fuel production.

Due to the high operating temperatures, plasma is very effective at vaporizing very difficult waste materials. Plasma gasification is also more robust than other gasification systems which are closely engineered to match the feed stocks being used. Many forms of gasifiers are used for coal and biomass, but plasma systems are unique in their ability to mix and match feed stocks, and even vaporize raw municipal waste, which may include metals, glass and electronics. Tires, medical waste, petroleum refinery wastes, low grade coal, railroad ties and phone poles are all examples of materials that are currently considered toxic and difficult to dispose of and yet are ideal fuels for plasma gasification and can be used to produce clean energy.

All of the non-organic materials contained in the feed stock are melted and pour out of the bottom of the gasifier. This material is called slag, and cools into vitrified glass similar in appearance to obsidian. Slag is very stable and safe, due to its tightly bound molecular formations. It has been subject to many tests and easily passes EPA standards for leachability. Slag may be used as an aggregate in asphalt or concrete and may be subject to various value-added processes to separate metals and form bricks, tiles, or rock wool.

Waste Gasification Cleans Up the Environment

Waste gasification is good for the environment because it gives value to garbage and keeps it out of the landfills. Landfills produce significant amounts of methane, which is considered to be a potent greenhouse gas. Landfills produce toxic liquid leachate that must be collected to prevent contamination of groundwater and aquifers.

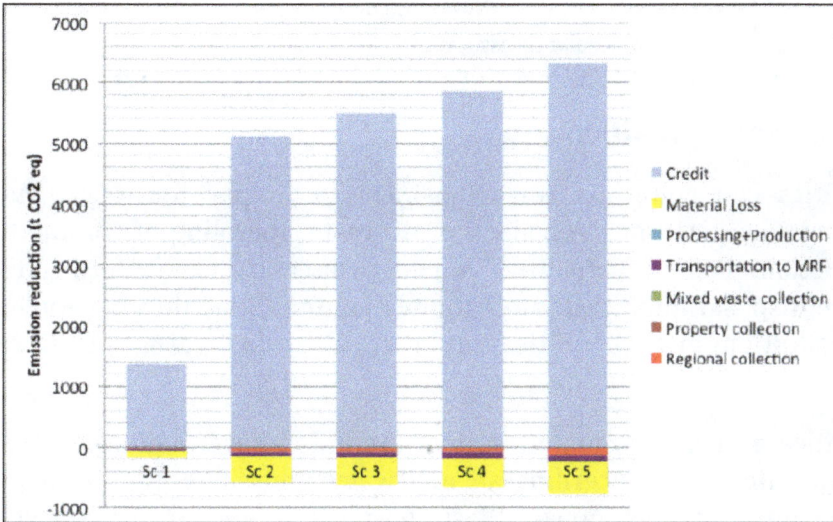

Gasification is not incineration and is a distinctly superior environmental solution compared to burning. The overall emissions of primary pollutants are very low from gasification. Gasification also does the best job of reducing overall greenhouse gas emissions, compared to other forms of waste management.

	Gasification	Incineration
Carbon Footprint	• -78,0000 MTCE per 1MM tons of MSW. • Sequestering possible.	• -18000 MTCE per 1MM tons of MSW. • No sequestration.
Air Emissions	• Minimal dioxins (.002 ng/m³ and furnas.	• Bad history of dioxin (.42 ng/m³) emissions.

	• Pre-combustion cleanup of syngas. • Similar to IGCC – Sox (1.2 ppmv), NOx (31ppmv).	• Post-combustion cleanup (scrubbing, filters, EP). • High SOx (9.3 ppmv), NOx (120 ppmv), particulates.
Ground Emissions	• Slag-safe and non-leachable. • 250:1 volume reduction.	• ~ 250 TPD ash per 1,000 TPD MSW, 4:1 reduction. • Ash is toxic, leachable and most often landfilled. • Concentrates of heavy metals, dioxins, chlorides.
Useful Products	• Net electricity ~ 900 kWh/ton. • Ethanol ~ 100 gallons/ton (in development). • Vitreous slag – useful for construction. • Recovered metals, sulfur.	• Net electricity ~ 550 kWh/ton. • Recycled metals.
Temperature	• Plasma at 1500 °C. • Sealed system – low oxygen. • Molecular disassociation.	• Fire at 850 °C. • Open air – excess oxygen.

Results of Waste Gasification

Municipalities can count on waste gasification to pay for itself. Many revenue streams emerge from the collection of waste, recycling of commodities, and fuel production. For municipalities, waste gasification can be utilized to transform waste disposal liabilities into valuable commodities that have value for the public. In addition, many liabilities can be avoided when garbage is diverted from landfills.

Municipalities must pay to maintain their landfills forever and are subject to regulations from the EPA concerning air and water emissions. By avoiding landfills, the municipality can save money. Waste gasification encourages robust recycling because commodity recyclables are far more valuable when sold than when used for fuel, and their removal improves the gasification operation and the quality of the output gas. Municipal solid waste (MSW) is charged a tipping fee at disposal that ranges from $30 a ton up to over $100 a ton for places like New York City where disposal is difficult. The output fuels earn revenues. The operator gets paid to take the waste and then gets paid again as the waste is processed and its value captured. Many products can be delivered from gasification. Liquid fuels, hydrogen and synthetic natural gas are all valuable products, but work is still being done to make their production from garbage profitable. The most immediate fuel product that can be delivered is electricity.

References

- Pizzi; K. L. Mittal (2003). Handbook of Adhesive Technology, Revised and Expanded (2, illustrated, revised ed.). CRC Press. P. 1036. ISBN 978-0824709860

- Plasma-activation: plasmaetch.com, Retrieved 27 May, 2020

- Evgeny V. Shun'ko & Veniamin V. Belkin. "Cleaning Properties of atomic oxygen excited to metastable state 2s22p4(1S0)". J. Appl. Phys. 102: 083304-1–14. Bibcode:2007JAP...102h3304s. Doi:10.1063/1.2794857

- Plasma-functionalization: plasmaetch.com, Retrieved 28 June, 2020

- Coburn, J. W.; Winters, Harold F. (1979-05-01). "Ion- and electron-assisted gas-surface chemistry—An important effect in plasma etching". Journal of Applied Physics. 50 (5): 3189–3196. Bibcode:1979JAP....50.3189C. Doi:10.1063/1.326355. ISSN 0021-8979

- What-is-plasma-electrolytic-oxidation-article: blog.keronite.com, Retrieved 29 July, 2020

- Koller, Albert. "The PPV Plasma Polymerization System: A New Technology for Functional Coatings on Plastics" (PDF). Balzers Ltd. Retrieved 17 March 2011

PERMISSIONS

All chapters in this book are published with permission under the Creative Commons Attribution Share Alike License or equivalent. Every chapter published in this book has been scrutinized by our experts. Their significance has been extensively debated. The topics covered herein carry significant information for a comprehensive understanding. They may even be implemented as practical applications or may be referred to as a beginning point for further studies.

We would like to thank the editorial team for lending their expertise to make the book truly unique. They have played a crucial role in the development of this book. Without their invaluable contributions this book wouldn't have been possible. They have made vital efforts to compile up to date information on the varied aspects of this subject to make this book a valuable addition to the collection of many professionals and students.

This book was conceptualized with the vision of imparting up-to-date and integrated information in this field. To ensure the same, a matchless editorial board was set up. Every individual on the board went through rigorous rounds of assessment to prove their worth. After which they invested a large part of their time researching and compiling the most relevant data for our readers.

The editorial board has been involved in producing this book since its inception. They have spent rigorous hours researching and exploring the diverse topics which have resulted in the successful publishing of this book. They have passed on their knowledge of decades through this book. To expedite this challenging task, the publisher supported the team at every step. A small team of assistant editors was also appointed to further simplify the editing procedure and attain best results for the readers.

Apart from the editorial board, the designing team has also invested a significant amount of their time in understanding the subject and creating the most relevant covers. They scrutinized every image to scout for the most suitable representation of the subject and create an appropriate cover for the book.

The publishing team has been an ardent support to the editorial, designing and production team. Their endless efforts to recruit the best for this project, has resulted in the accomplishment of this book. They are a veteran in the field of academics and their pool of knowledge is as vast as their experience in printing. Their expertise and guidance has proved useful at every step. Their uncompromising quality standards have made this book an exceptional effort. Their encouragement from time to time has been an inspiration for everyone.

The publisher and the editorial board hope that this book will prove to be a valuable piece of knowledge for students, practitioners and scholars across the globe.

INDEX

www.ingramcontent.com/pod-product-compliance
Lightning Source LLC
Chambersburg PA
CBHW061949190326
41458CB00009B/2827